YOUR KNOWLEDGE HAS VALUE

- We will publish your bachelor's and
  master's thesis, essays and papers

- Your own eBook and book -
  sold worldwide in all relevant shops

- Earn money with each sale

Upload your text at www.GRIN.com
and publish for free

Bibliographic information published by the German National Library:

The German National Library lists this publication in the National Bibliography;
detailed bibliographic data are available on the Internet at http://dnb.dnb.de .

Imprint:

Copyright © 2009 GRIN Verlag, Open Publishing GmbH
Print and binding: Books on Demand GmbH, Norderstedt Germany
ISBN: 9783668386808

This book at GRIN:

http://www.grin.com/en/e-book/340702/an-analysis-of-the-rice-market-integration-
in-uganda

Paul Waluube

An Analysis of the Rice Market Integration in Uganda

GRIN Publishing

AN ANALYSIS OF THE RICE MARKET INTEGRATION IN UGANDA

BY

WALUUBE PAUL
B. Sc. AGRICULTURE. (Hons)

A THESIS SUBMITTED TO THE SCHOOL OF GRADUATE STUDIES IN
PARTIAL FULFILLMENT OF THE REQUIREMENT FOR THE AWARD OF A
MASTERS DEGREE IN AGRIBUSINESS MANAGEMENT OF MAKERERE
UNIVERSITY

NOVEMBER 2009

DEDICATION

This thesis is dedicated to my parents, Mr. S. Waluube and Mrs. Alice Lovisa Nabirye who struggled to lay my foundation on whose sweat I have based to achieve higher strides.

To my dear wife, Winnie, and children; Humphrey, Penelope and Henry who were by my side through the thick & thin during the execution of this project.

ACKNOWLEDGMENT

I am highly indebted to my supervisors to Dr. B. Kiiza and Dr. Elepu Gabriel for their tireless supervision and guidance throughout this period of conducting research. Their guidance has been paramount to the completion of this postgraduate program. I wish to acknowledge the contribution of the entire staff of the Department of Agricultural Economics & Agribusiness, Faculty of Agriculture, Makerere University to the success of this research and postgraduate course in general.

Special acknowledgment goes to the Belgium Technical Cooperation, which sponsored my postgraduate program including research through its local scholarship grant program. I am thankful to I@Mak for their additional financial support towards this research.

I would like to express my thankfulness to my parents, family members especially Brothers Joel and John and Sisters Monica, Janet, Milly and Grace for their indispensable financial, material, and moral support particularly their prayers for my success.

To friends; Habajja Samuel, Samson Katengeza, Douglas, Fred Lugojja, Natanga Patrick, Kwizera Musaba, Robert Anyang, Dr. Isabirye Moses and Alice Amoding for their invaluable support and encouragements during this period.

Finally, I wish to acknowledge the assistance of the staff of International Food Policy Research Institute especially Dr. Todd Benson and David Luwandagga for their assistance in data re-organization from the *FoodNet* project documents.

ABSTRACT

It is crucial that formulation of market enhancing policies to improve performance hinges on a better understanding of the local market functioning. Comprehensive market performance is better understood by studying the extent of spatial market integration which in turn is affected by communication factors like telephony growth and extension services within the marketing system. Extension services are important in influencing levels of production and dissemination of trade information of both the inputs to and output from the enterprise. Unfortunately very little was known about how rice price transmission takes place. Given the importance of rice (as a food security crop & source of income) to the rural poor, many organizations have promoted it among rural farmers for improving their welfare, it was imperative that its price movements is studied to inform policy makers & implementers on the status of its spatial market integration.

The primary objective of this study was to evaluate the spatial market integration of rice markets under improving communication infrastructure particularly extension services (NAADS program) and telecommunication industry growth. This study specifically evaluated the degree of spatial market integration and strength of integration using bivariate analysis, Johansen Multiple test, and granger causality relationships. It also determined the impact of NAADS program and telecommunications on spatial rice market integration. Impulse response functions were generated to determine how long shocks in Kampala market can be eliminated in peripheral markets. Weekly prices data from sixteen markets from 2000 to 2006 were divided into three phases; full sample, pre – NAADS expansion and post NAADS program expansion to enable the study to capture the impact of these communication infrastructure.

Results from this study indicate that the markets are spatially integrated and with an incredible speed over the study period. Comparatively, there was more co integration and interdependence of rice markets in the post-NAADS expansion phase than the pre – NAADS expansion phase. A further analysis of the markets using multivariate analytical approach revealed a growing rice economic market constituting eight markets; Kampala, Jinja, Iganga, Lira and Rakai including Gulu, Mbale and Soroti markets with which growth began in the pre-expansion period. Overall the results reveal strong spatial market integration of the rice commodity in the post expanded phase of the communication infrastructure.

The study recommends the continuity of the NAADS program and improved telecommunications, continuous promotion of marketing associations and networking since the overall results depicted improved market integration. It also recommends the NAADS program to incorporate provision of timely agricultural and marketing information to strengthen the spatial integration of the rice markets. It also recommends related marketing studies to consider other structural determinant factors of market integration, and similar promoted enterprises to compare these findings.

Key terms: *NAADS program, telecommunications growth, rice, markets, prices, spatial market integration, co integration, Granger causality and impulse response functions.*

Table of Contents

ACRONYMS

AIC	–	Akaike Information Criterion
ADC	–	Agribusiness Development Centre
ADF	–	Augment Dickey Fuller test
APEP	–	Agricultural Production Enhancement Program
FAOSTAT	–	Food and Agriculture Organization Statistics
FEVD	–	Forecast Error Variance Decomposition
IDEA	–	Investment in Development Export Agriculture
IRF	–	Impulse Response Functions
NAADS	–	National Agricultural Advisory Services
VAR	–	Vector Auto Regression
VEC	–	Vector Error Correction Model
MAAIF	–	Ministry of Agriculture, Animal Industry & Fisheries
MFPED	–	Ministry Of Finance Planning & Economy Development
MTN	–	Mobile Telecom Network
NARS	–	National Agricultural Research Systems
SIC	-	Schwartz Information Criterion
RATIN	–	Regional Agricultural Trade Information network
WARDA	–	West African Rice Development Association

CHAPTER ONE

1.0   INTRODUCTION

1.1   The Ugandan Agricultural Sector

Agriculture is still the mainstay of Uganda's economy, accounting for 42% of GDP, over 85% of export earnings and providing about 80% of employment to the Ugandan population, 90% of who live in the rural areas (Anon, 2004). Food crops are predominant in the sector, contributing approximately 50% of agricultural GDP in 2003/4 with cash crops constituting 17% of this GDP (UBOS, 2004). FAO (2006) reports that food crops in Uganda are produced in two seasons, with estimated figures putting cereals production to 2.657 tones. With satisfactory food supplies, a large number of people are reported to have limited access to food due to low purchasing power.

1.2   Rice Production in Uganda

Rice growing in Uganda is mainly by smallholder farmers concentrated in the eastern and northern parts notably in the districts of Gulu, Iganga, Tororo, Mbale, Pallisa (Mukiibi, 2001), Lira, Kitgum, and Kumi (Imanywoha, 2001). Its production is in extensive swampy and high potential areas found around Lake Kyoga (Oryokot, 2001). The average land holdings range from 0.38 to .97 hectares (A.R. Ochollah *et. al., 1997)*. However, since 1992 total acreage of rice in Uganda has marginally increased (MAAIF, 2000). The total area under rice production increased from 50,002 ha in 1992 to 72,000ha in 2000 while the total production increased from 68,000 tons to 108,000 tons with a unit increase in yield from 1.4 tons to1.5 tons per ha as indicated in the following table;

Table 1: Production statistics of rice in Uganda

Year	Area planted (Hectares)	Output (tons)	Yield (tons/ha)
1992	50,002	70,000	1.4
1993	53,001	74,000	1.4
1994	55,001	77,000	1.4
1995	55,227	77,461	1.4
1996	58,000	81,000	1.4
1997	60,000	78,000	1.3
1998	64,000	90,000	1.4
1999	68,000	95,000	1.4
2000	72,000	108,000	1.5

Source: MAAIF Statistics Department, 2000

According to UBOS (2002), the area planted to rice in Uganda in 2001 was 76,000 hectares and the volume of paddy rice production was about 114,000 tones. This showed a significant increase in production compared to 1996 in which planted area was 58,000 hectares and yield to be 82,000 tones. This represented a percentage increase of 31% and 39% for area planted and rice production respectively.

1.2.1 Rice production trends

The rice crop in Uganda was introduced on a larger scale only in early 1960s and by these periods, the annual harvested acreage was just above 2,000ha. This added up to less than 3,000 Metric tones per annum. The land area cultivated under rice rose substantially in the 1970s due to government interventions; however, this was followed by a production stagnation reducing the area planted with the crop by about 1,000ha from 18,300 ha in the 1970s to 17,000 ha in the 80s. This was later followed by a minimal rise in the production from 18,000 tons to 22,100 tons (UBOS 2005, ADC/IDEA Project 2001, P.8). The Africa Rice Center (WARDA, 2007) reported Uganda's rice production between 2001 and 2005 to be at an average of 85.57 thousands of tones. This trend reveals that this crop enterprise is ever gaining importance through production and acreage.

1.2.2 Rice supply and demand

In Uganda, the domestic rice production can no longer meet the rising demands of especially all urban dwellers. Therefore, Uganda has become heavily dependent on imports of the rice food, making it one of her largest imports. Though rice production registered a steady growth between 1999 and 2003, averaging 114,000 MT, imports yet peaked during these years at more than 50,000 MT with an average of 41,000 MT arriving in the country each year. This implied that for Uganda to be able to meet the increasing demands, more than a quarter at times even a third, of the national rice supply had to be imported in the last five years as tabulated hereinafter. According to Jane Ininda, (2005) African rice consumption exceeds production and only 54 percent of sub-Saharan Africa rice consumption is supplied locally."

Table 2: Uganda's rice import demands against production from 1999 to 2003

Year	Production (Mt)	Imports (Mt)	Total supply (Mt)	% of imports/total supply
1999	95,000	39,741	134,741	29
2000	109,000	51,257	160,257	32
2001	114,000	22,225	136,225	16
2002	120,000	43,000	163,000	26
2003	132,000	48,925	180,925	27
Average	114,000	41,030	155,030	26%

Sources: Uganda Bureau of Statistics, 2005 and FAOSTAT, 2004b

According to statistics, while the national production growth averaged 6%, the import growth rate was ten times higher i.e. 62%.

Regionally, Tanzania is the largest rice producer averaging over 650,000 tones per annum of unshelled rice from 2000 to 2004 but it is also the largest consumer of this grain (table 3). In contrast, Kenya is the smaller producer and largest importer of rice in the region (APEP, 2006).

Table 3: Selected East Africa's Production and imports for rice from 2000 to 2004

Country	Production average last 5 yrs	Imports average last 5 yrs	Total consumption
Uganda	118,400	66,512	184,912
Tanzania	653,298	229,394	882,692
Kenya	46,430	240,382	286,812
Rwanda	24,464	17,030	41,494

Source: FAOSTAT, online

From the available statistics, all countries in the region are deficit markets that have over the years resulted in informal cross border trade amongst them. For instance, informal rice exports to Kenya from the region are pronounced. Tanzania's rice exports to Kenya had been increasing since 2000 and were worth UGX 13 million in 2004. This implies that a huge opportunity exists for the Uganda's rice industry in the region (RATIN, 2005).

1.3   Economic importance of rice

Among cereals, rice and wheat share equal importance as leading food sources for humankind (www.ggs.com/history/rice in human life). Rice is a staple food for nearly one-half of the world's population. In Africa, it is rice or wheat, followed by maize, yams and cassava that constitute the main food crops. In Uganda rice is increasingly becoming an important source of income for rural households in Uganda (Agricultural Policy Committee, 1998) and it is the fourth important crop after maize, finger millet and sorghum (Oryokot, 2001). It is second to none in economic return to peasant farmers on the basis of labour per annum per day per Ha and the main consumers of Uganda's rice are mainly the urban populations. This was revealed further by the national household survey conducted UBOS in 1999/2000 (UBOS, 2002).

1.4   Rice consumption in Uganda

Rice consumption in Africa has a high-income elasticity, and its demand is linked to urbanization and economic growth (African Crops, 2004 & WARDA, 2005c). In Uganda the urban population more than doubled between 1980 and 1991 (Table 4), reaching almost two million persons while urban population rose by 11% (UBOS, 2002).

Table 4: Trends of urbanization in Uganda (1969 – 2002)

Index	1969	1980	1991	2002
Total population	9,535,051	12,636,179	16,671,705	24,442,084
Urban population	634,952	938,287	1,889,622	2,999,387
Urbanization level (%)	6.6	7.4	11.3	12.3

Source: Uganda Bureau of Statistics, 2002

1.5   Problem statement

After the launch of the Plan for Modernization of Agriculture, PMA in 2000 (MAAIF/MFPED, 2000), improved dissemination of agricultural information was embarked on by introducing National Agricultural Advisory Services (NAADS) program and liberalization of telecommunication infrastructure which had started in the mid 90's. Among the important agricultural information are the prices of agricultural commodities and products. Such information is easily transmitted through an efficient extension service system like NAADS and telephony usage by marketers. Therefore an improvement in telecommunications and expansion of NAADS program is supposed to bring out strong spatial market integration.

Attempts to understand the impact of such interventions that have lasted for almost ten years are important to inform stakeholders on policy impact and related issues. Available information reveals rice as the fourth important crop after maize, finger millet and sorghum (Oryokot, 2001) which enterprise importance has been increasing as source of income for rural poor households (APC, 1998) that can alleviate poverty among them and a food security crop (Odogola R. Wilfred, 2006). A study of the rice market performance in form of market integration plays a significant role as one way of achieving such an objective.

The Uganda government has tried to improve the marketing infrastructure since a poor infrastructure negatively affects market integration as reported by Goletti & Christinas-Tsigas, (1995). For instance with low telephone usage fewer marketers can easily share price information in time, like wise with NAADS program operating in a few districts its contribution to price dissemination, promotion of production is reduced. The government introduced the NAADS program, alongside the liberalized telecommunication industry to sharing of agricultural information including prices of products or commodities. In the rice industry specific efforts included promotion of rice production and marketing through NAADS program, Vice President's Poverty eradication projects where rice seed was widely distributed as an in-kind credit (Yoko Kijima, 2008).

Various factors contribute to the efficiency of a commodity market. Among them are marketing infrastructure (like rural road network, telecommunications), the food policy, the degree of dissimilarity in production in the markets. The more diverse the markets are in terms of rice production, the more incentive it is for them to trade together, hence the higher the likeliness for spatial integration (Goletti, Ahmed & Farid, 1995). On this foundation, spatial market integration was used as surrogate for improved market efficiency.

Market integration is a precondition for effective reform in many of the former centrally planned economies (Baulch, 1997). Therefore, when integrated markets are identified carefully through study, a regional balance between surplus and deficit regions (Goletti *et .al.*, 1995; Kherallah, 2000) can be planned whereby closely integrated are grouped together and designing market liberation, credit and transportation policies (Goletti, Ahmed and Farid, 1995). This avoids duplication of government interventions, and aid in making forecasts.

With such a study, the various institutions can be equipped with information to wisely inform the agents in the rice industry on the best ways of objectively improving or profitably benefit from the rice enterprise, contributing to poverty reduction. Because the study can measure the extent to which demand and supply shocks in one location are transmitted to other locations (Negassa *et al.,* 2003) and respective price transmission signal durations, it helps policy makers to know the response of markets to various shocks and the adjustment periods necessary for a given policy success. This is important in informing policy makers on to not only design but also improve food security initiatives and poverty reduction interventions through agriculture.

Unfortunately, due to almost no research on rice markets and marketing, there was limited information on the spatial market integration. Previous studies on rice enterprise in Uganda focused mostly on finding ways of increasing productivity by analyzing productivity factors. The need to have studies to generate information on the market integration across the markets in the country was glowing and a detailed study on the degree of spatial integration of the rice markets was imperative. There was a further need to attempt to review the extent of this spatial integration following the commencement of the NAADS program under the increasing telecommunication improvement. Notably, it would depict the NAADS program and other marketing infrastructure interventions influence contributions and progress in enhancing market integration. In so doing, the study results contribute to strategic and tactical decision making in the agricultural sector and provide technical ground for further re-engineering their strategies of implementation.

1.6 General Objectives

The main objective of this study was to examine spatial rice market integration in Uganda following the growth and expansion of the telephony and road network in the country.

1.6.1 Specific objectives

The specific objectives were:

a) To determine the degree of spatial rice market   integration in Uganda.

a) To determine causal relationships among the rice markets in Uganda

a) To determine the dynamic adjustments of rice markets to shocks in the long-run

1.6.2 Hypotheses of the Study

1) Spatial integration of the rice markets has improved over time.

1.7 Scope of the study

The study was aimed at reviewing the performance of the marketing system of the rice sub sector. With an intention of obtaining the entire country rice-market integration, sixteen districts were purposively selected from all regions namely eastern, northern, and western and central/Buganda regions. To achieve the objectives, secondary weekly rice prices data for use were collected from *FoodNet* project of *IITA*.

CHAPTER TWO

2.0   LITERATURE REVIEW

2.1   Market linkage and integration

2.1.1 The concept of Integration

In a market economy, the marketing system serves as mechanism to transmit to market participants' information that is useful in decision making. Transparent, accurate and timely price signals play a leading role in shaping the conduct and performance of an efficient marketing system. In a competitive market environment, the pricing mechanism enables the transmission of orders and gives directions that determine the flow of market activities. These signals guide and regulate production, consumption and marketing decisions over time, form and place (Kohls and Uhl, 1990). Identifying the causes of differences in prices in interregional or spatial markets has therefore become an important economic analytical tool used to provide a better understanding of markets including agricultural product markets. The next section explores this concept.

In his explanation using population data as an example, Clive W. J. Granger (2004) explained that market price data could be available annually or less frequently. He explained that when such data series is smooth, moving with local trends or with long swings but swings are not regular, then that data that is not smooth as required by standard statistical procedure is said to be integrated or nonstationary. The difference between a pair of integrated series can be stationary and this property is known as Co integration. For co integration, a pair of integrated or smooth series must have the property that a linear combination of them is stationary. Most pairs of integrated series will not have the property of stationarity, so that co integration should be considered as a surprise when it occurs.

2.1.2 Market integration

Integrated markets are those whose prices for a given commodity or commodities in different localities have a stable long run relationship (Goletti & Christinas-Tsigas, 1995). Spatial market integration refers to co-movement of prices, and more generally, to the smooth transmission of price signals and information across spatially separated markets (Goletti, Ahmed & Farid, 1995). This phenomenon is also explained by correlations of price series of different markets in which it is ideological that integrated markets exhibit prices that move together although in some cases, parallel movements in prices can occur for several other reasons other than the integration of markets.

Market integration can also be defined as a measure of the extent to which demand and supply shocks in one location are transmitted to other locations (Negassa *et al.*, 2003). Barrett (1996) distinguished market integration into (i) vertical market integration involving different stages in marketing and processing channels, spatial integration relating to spatially distinct markets, and (ii) inter-temporal market integration which refers to arbitrage across periods. This study intends to examine spatial market integration.

Gonzalez-Rivera and Helfand (2001) noted that while there is a general agreement that market integration somehow relates to the flow of goods & information across space, time and form, getting a widely acceptable definition of integration proved to be elusive. They therefore proposed another definition that emphasizes trade and information. Accordingly, Gonzalez-Rivera *et. al.* (2001) defined integrated market as a market with a set of locations that share both the same traded commodity and same long run information. In the co integration framework, the second condition requires the existence of one and only one co integrating factor common to all price series within the market. These criteria are important in identifying the sets of locations that are spatially linked directly or indirectly by trade. Under this co integration study, market integration analysis was extended from bivariate to multivariate framework, which looks at market integration in terms of a continuum of degrees of integration (Shahidur, 2002). The underlying idea is that for a set of spatially separated locations, not all locations belong to the same economic market and among all those locations that belong to the same market, some are more integrated than others.

One of the main consequences of poor market integration is the difficulty with which information and trade flows occur in spatially separated markets (Goletti & Chritinas – Tsigas, 1995). However, knowledge on market integration is important for grouping closely integrated and designing market liberation or price stabilization, credit and transportation policies (Goletti, Ahmed and Farid, 1995), to avoid duplication of government interventions, and making forecasts to guide other stakeholders. This had been emphasized earlier by Goodwin and Schroeder (1991) that markets that are not integrated may convey inaccurate price information that distort producer marketing decisions and contribute to inefficient product movements resulting in poor market performance. Market integration knowledge is a precondition for effective reform in many of the former centrally planned economies (Baulch, 1997).

Proper market linkage is one way of improving performance of product markets to benefit farmers and traders (arbitragers). For instance if there is an uneven distribution of food supplies in the country, assurance of regional balance between deficit and surplus regions (Goletti *et. al.*, 1995; Kherallah *et. al.*, 2000 and Istiqomah *et. al.*, 2005) is achievable through proper knowledge of markets generated through market integration studies. Therefore, measurement of market integration can be viewed as a basic tool for an understanding of how markets work (Ravallion, 1986). It is upon such information that important implications for economic development (Gonzalez G. Rivera and Helfand, 2001) can be drawn for a country.

2.1.3 Measures of Market integration

Goletti and Christinas –Tsigas (1995) explained the various methods of measuring markets integration namely; descriptive analysis of market network, use of correlation and co-integration tests of the prices of given commodities in different markets, dynamic adjustment and composite indices. Use of correlation coefficients is based on comparison of correlation of prices of markets. This is because it is assumed that integrated markets have prices that move together over time.

The study of market integration based on correlation coefficients, however, has faced criticism because according to Goletti *et. al.*, (1995) it does not address factors like price inflation, seasonality, population growth and procurement policy for government managed procurements of the commodity whose prices are considered. This is why in some studies it is not applied. Criticisms are raised because of its failure to address the influences of the above factors particularly in agricultural commodity markets where peak and deficit seasons often occur at the same time results in spurious correlations. This is why such specious correlations are eliminated by considering price differences instead of price levels in computations of correlation coefficients. Spurious correlations are not the only problems in measuring market price correlations; other serious problems are related to the non stationary nature of prices, and the failure to determine the direction of integration among the markets.

Use of market network analysis is a good method of studying integration but necessary data for example on trade flows and marketing costs is not readily available and tends to be unreliable.

In some studies, understanding market integration is a sequence of studies for instance bivariate analysis of integration is conducted to identify existence of equilibrium relationships between market pairs over both the medium and long term (K. Weber and D. Lee, 2006). Multivariate co integration is then conducted to establish the boundaries of the continuous market, which is also called the economic market (Gonzales Rivera and Helfand, 2001). In the final study, markets with stable long run relationships are the ones that are subjected to impulse response analysis to draw insight into the speed at which such markets react to deviations from the equilibrium to restore the long run relations.

2.1.4 Empirical Studies of market integration

Various authors and researchers have used the co integration analysis in the study of market integration. Others extend their analyses by conducting the effect of shocks in markets to the rest of the markets through impulse response analysis using impulse response functions. This sub section reviews some studies that have applied the concept of co integration with the aim to compare various ways the concept is used in relation to spatial market analysis.

In their paper on regional cattle markets, Goodwin & Schroeder (1991) aimed at evaluating the co integration and spatial linkages among 11 regional cattle slaughter markets by determining the effect of several market characteristics on the co integration of the markets. On running several spatial price relationship tests, they found that very many markets were not co integrated between 1980 and 1987. They concluded by suggesting that trade and information costs would decrease and packers could coordinate the price behavior across regions as the market concentration increases. Based on the unexpected results obtained, it was argued based on Ardeni (1989) that conventional approaches of testing for spatial integration had misrepresentation or ignorance of time series properties of regional price data.

Goletti, Ahmed & Farid (1995) studied 64 coarse rice markets at district level in Bangladesh with an aim of determining the main factors responsible for market integration. They hypothesized that marketing infrastructure, volatility of government intervention and degree of self sufficiency are the major determinants of market integration. They addressed this hypothesis by application of a two-stage approach in which four measures of market integration were used namely; correlation coefficients, co integration coefficients to capture the long term relation among the price series, long term multipliers that express the cumulative response of one market to price shocks originating from another market. They lastly conducted the fourth measure by estimating the speed of adjustment of markets to the long term multipliers.

Results from their study indicate that different measures respond differently to market integration structures, and the level of rice market integration for the period under study was moderate. The results further revealed that distance negatively affects market integration while production shocks positively affect market integration of the given markets. Once distance between markets under consideration is less than the critical distance of 250 kilometers from each other, the level of market integration is higher. It is noted that to get results that are more meaningful price and structural data that cover longer period are necessary.

While studying South African apple markets within the fresh produce markets, David I. Uchezuba (2005) used various methods to measure market integration of apples on South African fresh produce markets. He used ADF and Johansen tests to investigate long run relations in the markets. He then fitted threshold vector error correction models to investigate the dynamic relationship among apple markets and applied regime switching and impulse response functions to evaluate switches among regimes and responses to shocks by various markets. Major findings in his study were that market integration in the markets was evident with both unidirectional and bidirectional causality among selected markets. He discovered that there was sufficient arbitrage among the South African fresh produce markets, which is consistent with market integration.

In Uganda, increasing need for regional agricultural production specialization has been agitated for since early 2000 under the Plan for Modernization of Agriculture. This was done through zoning the whole country into twelve agro-ecological zones (NARS, 2002). Under such circumstances, some enterprises are assumed to perform better in some regions than in others. With such policy arrangements, market integration studies are very important to inform policy on which way to advise farmers on regional basis. A similar scenario study was conducted by K. Weber and D. Lee (2006) in which they used market integration and response analysis to examine whether by comparative advantage, markets in the U.S. and Mexico adhered to the principles of specialization accordingly.

In their study, ten rice markets in U.S. and Mexico were analyzed by examining market integration through price convergence and co-movement through a sequential process. Through use of bivariate and multivariate analyses, they established the boundary of the continuous market and applied impulse response analysis techniques to get insight on how the speed of deviations from the equilibrium could be corrected. Similar methods including Johansen sequential testing procedure were used by Titus O. Awokuse (2007) on the China's rice markets. Their results were that under the long run equilibrium most of the Mexican markets are bound to the U.S. markets. Awokuse (2007) study further revealed that the U.S. and Chinese rice grain markets were co integrated with continuity. This means that some markets may be continuously co integrated while others are segmented from the continuous integration boundary.

Theingi Myint and Siegfried Bauer (2005) used co integration method for integration analysis of Myanmar rice market. In their explanation, they stated that integration means that if two variable series say p_{it} and p_{jt} are each non stationary in levels but stationary in first differences, that is $P_{it} \sim I(1)$ and $P_{jt} \sim I(1)$, there exists a linear combination between the two series that is stationary. They emphasized that in an efficient market system, prices move together in that trade takes place if prices in the importing region equals prices in the exporting region plus the unit transport cost incurred by moving between the two (Ravallion, 1986).

The result of efficient trade and arbitrage activities is that prices at different market places cohere (P.J. Dawson and P.K Dey, 2002). This is why co integration analysis allows verifying whether markets have efficient trade or not. In their study, co integration analysis was conducted using weekly prices of pairs of markets Mandalay-Yangon and Yangon – Mandalay for one type of rice product (pawsan). The most striking feature of the results is that the rice prices were highly co integrated between Yangon and Mandalay market during the 2001 and 2004.

To provide an insight understanding of the effect of trade liberalization on the rice market in Java provinces, Indonesia, Istiqomah, Manfred Zeller and Stephan von Cramon – Taubadel

(2005) applied the multivariate Johansen maximum likelihood method to comparatively analyze the volatility and degree of integration in two situations i.e. in before and after –trade liberalization. In their method, it is evident that co integration analysis is conducted after confirming that price series are integrated to first order in the first differences. The results of this study indicated that among the five markets studied there were four co integrating vectors among prices series of five markets on Java in the pre liberalization and only two co integrating vectors in the post liberalization period. They concluded that the less integration could have been due to delayed response of markets to the new policy of liberalization.

Vietnam is one of the world's largest rice producing and supplier countries but at the same time, the country has been undergoing various reforms. These included rice market liberalization after 1989 and with the various initiatives involved in these reforms, better market performance was expected. To evaluate this transition policy, Clemens Lutz *et. al.,* (2006) used the Johansen maximum likelihood estimators to test for multiple co integrating vectors to assess the long run market integration in the country. The results of this study showed that all market places under study were integrated in the long run. The prices in Ho Chi Minh (HCM) city and the markets in the Mekong River Delta were strongly correlated even in their short run. On application of the VECM, all markets, except for the export price series, reacted strongly to deviations from at least one of the long run co integrating relations.

In their study of the food market integration with focus on the Vietnamese paddy market, Le Dang Trung *et. al,* (2007), used transfer costs to generalize the well known model of spatial market integration due to Ravallion to allow for the possibility of threshold effects. On application of the unrestricted version of this model using monthly paddy prices for eight markets between 1993 and 2006, a weak evidence of market integration was found in the north and south of Vietnam without any threshold effects. Their further analysis of paddy markets, revealed threshold effects and stronger forms of spatial market integration within the north and within the south of Vietnam.

Shahidur Rashid (2004) studying Uganda's maize market integration in the post liberalization period discussed the various groups of co integration analyses. These included Ravallion's radial market integration model in a co integration framework. The test for co integration is carried out as a test for unit roots on the saved residuals in which case if residuals are stationary, the markets will be co integrated. The second group of studies is one that uses Johansen's multivariate co integration method to draw inference about the strength of integration among regional markets. The idea behind this method comes from Stock and Watson (1987) who demonstrated that if a set of n economic variables is co integrated with exactly n-1 co integrating vectors, these variables must therefore share a common trend. Shahidur pointed out that full integration of markets requires exactly n-1 co integrating vectors and therefore any number that is less than n-1 implies weak integration. In this study, the results indicated that markets that were not integrated in the earlier period of liberalization became strongly integrated in the subsequent years.

In Brazil, rice is one of the major cereals produced in the country but this production is concentrated in small number of states. Gonzalez-Rivera *et. al.*, (2001) used this Brazilian rice market to illustrate that the multivariate approach is better than bivariate analysis in studying market integration. To improve on the knowledge of market integration two features namely searching for boundaries of the continuous market and use of persistence profiles to study the degrees of integration of locations that belong to the market were introduced through his study. This means that even among the integration the levels of integration differ by varying degrees. To get the continuous market, they proposed a sequential procedure based on Johansen (1988, 1991) to search for the single common factor.

In a continuous market, one and only one integrating factor must be common to all the price series. Given n-locations in a market, there must be n-1 co integrating vectors. When normalized, the n-1 co integrating vectors with respect to given a location, we find that all the locations were integrated pair-wise. Under this arrangement, it is very difficult to determine which locations belong to the same market using the bivariate approach because of all the $n(n-1)/2$ pair wise combinations only n-1 combinations would be relevant. Under this circumstance, it becomes complicated and gives inconclusive results contrary to the multivariate analysis. Bivariate analysis limits each equation in the VECM to only one error correction term and lags from only two locations considered. This grossly miss-specifies the model for some special market structures. The integrating factor is eliminated when the co integrating relation is estimated yet a common long run trend that gives rise to co integration has to be found.

Gonzalez-Rivera *et. al.*, (2001) stressed that to find which locations belong to the same market; we begin with a maximum set of locations, n, and testing for n -1 co integrating vectors. This is done by performing Johansen Likelihood ratio test based on the trace statistic. If the number of co integrating vectors is less than n -1, there is need to identify those locations that must be removed from the system by performing a sequential procedure. The procedure is started with the core set of m (m<n) and test for the number co integrating vectors, if the number is equal to m – 1 co integrating vectors, then we add another location. With m+1 locations, either the new one shares a common trend with the previous m locations or it does not. It is mentioned that in the first case, we should find m co integrating vectors, in the second we should continue to find m -1 co integrating vectors thus adding a second common trend to the m+1 co integrating vectors. If one common trend is found, the procedure is repeated through addition of another location at a time. In one of their investigations, they found that fifteen states belonged to the same economic market while four did not appear to belong to this market.

In 1998, structural changes took place in the grain/oilseed industry in the U.S. and various impacts had taken place. These were structural changes mostly on the pricing and linkages in the east and North Missouri prior to and following the January 1998 merger of Archer Daniel's Midland (ADM) and Quincy Grain Company and opening of the producer- owned ethanol plant in Macon, Missouri in 2000. These were structural changes observed i.e. mergers raised a need to find out whether the pricing pattern and linkages were altered as exhibited by the level of integration between the selected Missouri soybean markets. The hypothesis for this study was that the observed structural changes altered the pricing patterns and linkages as exhibited by the level of co integration between Missouri corn markets. In the methods, a three tier technique was used to analyzed the ADM, Quincy and the opening of the NEMO ethanol plant in 2000 on the price linkages and pricing patterns in Kansas City, Macon, Hannibal and St. Louis Missouri corn and soybean markets.

Using primary level local elevator corn and soybean prices, in one of their tier of analysis using impulse response functions, having found out that the market pairs were co integrated, the IRF were created using cholesky decomposition and converted the VAR models into moving average   representation of the inter-market   relationships. Impulse Response Functions were examined to determine how quickly prices in one location adapted to shocks in prices at another location prior to and following structural changes for soybean and corn markets (Jason R.V. Franken and Joel L. Parcell, 2003).

The graphical representations of impulse response functions showed that a one standard deviation shock in Hannibal soybean market requires 50 days for Kansas City to respond to this shock. The graph showed that the response was quicker after the merger. The same response period was observed in St. Louis after a one standard deviation in Hannibal soybean market. However, in both pre and post merger period the response is complete by around 50 days. For Macon, it takes 40 days following a similar shock in Hannibal soybean market. On the other side, Macon corn market responded to a one standard shock in Kansas City corn market by a small margin. The response of St. Louis corn market was completed after 50 days following a one standard deviation shock in Kansas City market.

2.2. Lag order determination Information Criteria

The idea of imposing a penalty for adding regressors to the model has been carried out further mathematically as follows:

$$(6) \qquad \ln AIC = \left(\frac{2k}{n}\right) + \ln\left(\frac{RSS}{n}\right)$$

Where ln = natural logarithm, 2k/n = penalty factor. n = number of observations. This AIC criterion is advantageous in both in-sample and of-sample forecasting performance of a regression model. Although the AIC has the above advantage, it at the same time only measures the discrepancy between the given model and the true model. The SBIC and HQIC on the other hand provide consistency of the true lag order, ρ (Gujarat, 2003). Lutz *et. al.*, (2006) noted that co integration tests are quite sensitive to number of lags to be included in the models. They recommended use of both AIC and SBIC criteria to select the suitable lag lengths where we begin with the maximum number of lags and then test whether the number can be shortened. J. Gonzalo and J. Pitarakis (2000) pointed out AIC as the best lag length estimator for large samples in co integration analyses.

CHAPTER THREE

3.0   METHODOLOGY

This chapter gives an overview of the study area, how the collection and processing of data were done before running the statistical tests. In this chapter the methodology used in measuring the price transmission, relationships and co integration are also discussed. The various statistical tests to be performed are discussed.

3.1   Field Methods

3.1.1 Study area

This study focused on 16 major rice markets in Uganda. Weekly data on rice prices were obtained from *FoodNet* for the period between 2000 and 2006. The markets are spaced across the whole country and they are; Owino-Kampala, Jinja, Iganga, Tororo, Mbale, Soroti, Arua, Gulu, Lira, Luwero, Masindi, Masaka, Rakai, Mbarara, Kabale and Kasese market. Some of these markets are located in major rice producing areas for instance Lira, Tororo, Gulu, Mbale & Iganga. The major consumer market was Owino -Kampala. These markets were subdivided into regional markets; northern markets with Gulu as the major regional market; Mbale as regional market of eastern rice markets; Mbarara as the major regional market for western markets. Owino market in Kampala represented both the central region and national dominant market.

3.1.2   Data Collection and handling

The main research instrument used was secondary data (from 2000 to 2006) obtained for all markets where *FoodNet* project used to collect rice prices on weekly basis. After raw data collection, each data series was cleaned and reformatted by means of adjusting the prices. After complete data entry, data were then crosschecked for validity and consistency using the MS Excel and STATA. Missing values were approximated by linear interpolation where there were one to three missing values. In case of more than three missing values, prices from nearby market were placed for missing values because it is supposed under spatial arbitrage theory that prices of the same commodity in adjacent markets tend to move in unison and do not divert from each other (Tomek & Robison, 1990). Analysis of the data was mostly performed using the STATA 9 program. The 0.05 level of significance for various statistical tests was used to test the study hypothesis.

The data were then set to have time series properties and transformed by two major transformations namely natural log and first difference transformations. The cleaned data was arbitrarily divided into three data sets i.e. full sample (2000 – 2006), first phase (2000- 2003) and second phase (2004 – 2006). By the end of 2001, the telephony (mobile telecommunication) and road network had just started expansion, while by September 2006; the number of telephone subscribers had greatly grown to over two millions up from under two hundred thousand in 2001 along improved road network in the country.

This segregation of data was intentionally done to enable the study track down the relative importance of telephony growth and in general the marketing infrastructure like road network in the improvement of spatial rice market integration in Uganda over this period.

3.2   Analytical Methods

3.2.1   The concept of market integration

The concept of market integration states that if a series Pit (rice price in market i at time t) is non stationary but its first difference is stationary, then it is said to be integrated of order one or simply integrated, and can be represented as $p_{it} \approx I(1)$. Otherwise, if p_{it} is stationary in levels it is said to be integrated of order zero and denoted as $p_{it} \approx I(0)$. If two series p_{it} and p_{jt} are both integrated to the first order, the linear combination of series i.e. $p_{it} - a - p_{jt} = \varepsilon_t$ is also integrated to the first order. It is possible for the error term to be integrated to zero order or stationary. This happens when the trends in p_{it} and p_{jt} are co integrated with b as the co-integrating parameter of coefficient. In general a pair of series p_{it} and p_{jt} is said to be co integrated if they are individually $I(d)$, where d is the order of integration and a linear combination of them, $\varepsilon_t = p_{it} - a - p_{jt}$ will be stationary i.e. $I(0)$ (---------------).

Measuring market integration is still a matter of considerable debate by concept and empirical approach. Uchezuba (2005) notes that to determine whether the arbitrage conditions are met requires data on prices, trade flows between markets and transfer costs with these transactions. In practice, it is only price information or data that is readily available because for instance transfer costs are not easily captured except through inclusion in the prices figures. Therefore, empirical market integration tests concentrate on analysis of prices of given commodity markets and this approach does not reveal whether there are trade flows among markets due to price differentials. Barrett (1996) notes that co-movement of prices has thus become synonymous to market integration. Spatial market integration refers to co movement of prices and more generally, to the smooth transmission of price signals and information across spatially separated markets.

Several approaches exist in literature to testing spatial market integration all relying on market prices to examine the concept of spatial arbitrage. Dawson and Dey (2002) list four econometric approaches of measuring market integration as identified by Baulch (1997) namely; the Law of One Price, the Ravallion model, Granger causality and co integration.

On the basis of the fact that on price information was collected by Foodnet from private traders in the study markets, the study tests the existence of co movement and price relationships among these markets using the co integration analyses.

3.2.1 Diagnostics and testing

The major objective of the research was to investigate the market integration state in the rice markets. Nevertheless, one major problem with integration studies is that most of the business and economic time series are never stationary. The literature suggests several approaches to testing spatial market integration using market prices to examine the concept of spatial arbitrage. Therefore, based on the principal objective, the product prices were converted into an Autoregressive model, and diagnostic checks for stationarity performed followed by series of tests as explained hereinafter.

Statistical properties of the Data

The first step in analyzing data for purposes of market integration is to check the statistical properties of the time series data. This is necessary to determine the nature transformations to be performed before further analyses.

Unit root tests

The standard check performed to test the properties of individual price series is the Augmented Dickey Fuller (ADF) test (Shahidur, 2002; Uchezuba, 2005). The ADF test is used to test the null hypothesis that a given price series p_t is non stationary against the alternative hypothesis that p_t is stationary by calculating a test statistic t for β=0 in the equation (1) assuming a random walk process:

(1) $\quad p_t = \alpha + \beta p_{t-1} + \varepsilon_t$

Where p_t and p_{t-1} are prices of rice at time t and t-1, β is the coefficient of lagged rice prices and ε_t is the random error and α is a constant drift. A null hypothesis means that the difference between the current and lagged prices of rice product in a given market is nil or insignificant. When a time series is non stationary, regression in levels often displays first order serial correlation and results in spurious regression results (Vinuya, 2007). A time series is said to be stationary if its mean fluctuates around a constant long-run mean and the variance is fixed.   If a time series is non stationary, it is said to follow a random walk process when $\beta \geq 1.0$ and it can be tested for the presence or absence of the random – walk process following relationship:

(2) $\quad p_t = \beta p_{t-1} + \mu_t \quad -1 \leq \beta \leq 1$

Where μ_t is a residual error term.

If the observed ADF test statistic is less than the critical values, then the p_t will be stationary. When p_t is non stationary, the next step is to determine whether this series is stationary at level (zero order) or not. If p_t is not stationary at level, it may be stationary at first difference of this p_t series. The differenced price series can be obtained by taking the first difference or simply differentiation of equation (2) through manipulation by subtracting p_{t-1} from both sides, so that we obtain

(3) $\quad p_t - p_{t-1} = \beta p_{t-1} - p_{t-1} + \mu_t$

And by factorization, equation (3) can be written as follows:

(4) $\quad \Delta p_t = \delta p_{t-1} + \mu_t$

Where $\delta = (\beta - 1)$ and $\Delta = p_t - p_{t-1}$, is the difference operator.

To test for stationarity in a in the differenced time series Δp_t in consideration, the null hypothesis is that $\delta = 0$ so that $\beta = 1$, in such a case equation (4) will have a unit root. Under this condition, the time series in consideration will be non stationary. Therefore in testing for stationarity, the first differences of p_t are regressed on p_{t-1} to see if the estimated slope coefficient in the regression is zero or not. This regression test to determine this slope through application of ordinary least squares (OLS) is what is termed as the Augmented Dickey Fuller (ADF) test. The regression is expressed as in equation (5) (Theinghi Myint and Siegfried Bauer, 2005).

$$(5) \quad \Delta P_t = \alpha + \beta P_{t-1} + \gamma t + \sum_{k=2}^{n} \delta \Delta P_{t-k} + \xi_t$$

Where $\Delta P_t = P_t - P_{t-1}$; $\Delta P_{t-k} = P_{t-k} - P_{t-k-1}$; k=2,3,……n, P_t is the price at time t; α, β, γ and δ_k are parameters to be estimated and ξ_t is the error term. K = number of lags to be included in the regression equation. The null hypothesis of a unit root is $H_0: \beta = 0$ in equation (5). The regression can be run with or without a time trend depending on the nature of the price series (Dawson and Dey, 2002). According to Bopape (2002), the trend is only included to rule out the possibility of non Stationarity not being due to a deterministic trend. The number of lags to include in the models was determined using AIC and SBIC with priority being given to AIC.

If the ADF test can be rejected for the null hypothesis (i.e. $H_0: \delta = \beta = 0$), it may be concluded that $P \sim I(1)$. Once the price series are confirmed to be integrated of one by rejecting the null hypothesis through ADF test of the differenced price series, the series are deemed possible candidates for co integration tests (Vinuya, 2007). The next step therefore is to test for co integration.

3.3   Testing for Co integration

3.3.1   Testing for long run co integration

Co integration analysis is concerned with providing information on the long run relationship of market prices of a commodity or commodities in a region or a country. The relationship can be Bivariate or multivariate among the price series. Prices in different markets move from time to time and their margins are subject to various shocks. When a long linear relationship exists between or among given price series, they are said to be co integrated (Engle and Granger, 1987). In an efficient market system, "trade takes place if prices in the importing region equals to prices in the exporting region plus the unit transport cost incurred in moving between the two" (Ravallion, 1986).   The result of efficient trade and arbitrage activities is that prices at different market places cohere (Myint & Bauer, 2005).   If we consider two economic time series p_{it} and p_{jt} which are both integrated to the first order, a regression of one series onto the other produces a linear combination of series i.e. explained by a long run relationship or co integrating coefficient β when the error term to be integrated to zero order or stationary. This happens when the trends in p_{it} and p_{jt} are co integrated with β as the co-integrating parameter of coefficient. The first step in testing for co integration is to run a co integrating regression of one price series p_{it} integrated to first order onto another integrated price series p_{jt}. A linear combination of two I(1) price series is non stationary in the short run but it may be stationary in the long run (Balke and Fomby, 1997). Engle and Granger (1987) define a dynamic long run relationship between two price series by the expression below;

$$(7) \qquad p_{it} = \alpha + \beta p_{jt} + \mu_t$$

Where p_{it} is the price of rice at market *i* at time *t* and p_{jt} is the price of rice at market *j* at time *t*, *α* is a constant, *β* is the coefficient of regression and μ_t is a residual error term. This test in equation (4) is the test for Law of One Price and it is only valid if the null hypothesis of no co integration is rejected (i.e. β=1). According to Dawson & Dey (2002) in perfect market integration, the null of perfect market integration is tested where price change in one market leads to an equivalent price change in another; imperfect market integration occurs where the price changes are not strictly proportional.

Clemens Lutz *et. al.*, (2006) indicated that for numerous markets under study (like for this study), only those markets that significantly contribute to the relationship are included in the regression. A Dickey fuller test is carried out on the residuals to verify whether the relationship is stationary. The relationship must satisfy the generalization of the Law of One Price.

New developments in time-series econometrics, suggest that if the price series are non-stationary, normal inference is not valid on the parameters and results from equation (7) are spurious. However, if the price series are integrated of the same order, then the series can be used to test for co integration using the Johansen vector autoregression (VAR) method.

Co integration implies that there is a linear long-run relationship between or among price series in spatially separated markets, and is interpreted as a test that $\beta \neq 0$. If $\beta \neq 0$, then the price series are co integrated and a long-run equilibrium relationship exists among the price series, and hence there exists a co integration vector (1, -β). Co integration tests for market integration are only tests of whether there is a statistically linear relationship between different data series (Asche *et. al.*, 1999) and tests for more general notion of equilibrium. The Johansen VAR-based procedure (Johansen, 1988) of testing co integration is the maximum likelihood procedure, which relies on the relationship between the rank of a matrix and its characteristic roots. The Johansen (trace) test detects the number of co integration vectors that exists between two or more integrated time series for a homogenous commodity. This test is used to test for the presence of a co integration vector between different price series if they are integrated of the same order. According to Granger representation theorem, this procedure is based on maximum likelihood estimation of the error correction model and each two-variable system is modeled as a vector auto regression (VAR) of P_t:

$$(8) \qquad \Delta P_t = \Pi P_{t-1} + \sum_{i=1}^{k+1} \Gamma_i \Delta P_{t-i} + \mu + \delta_t + \varepsilon_t$$

Where Π and Γ are n × n matrices of coefficients; $\Delta = (I - L)$ and L is a lag operator; k is lag length; α and δ are vectors of constants and trend coefficients respectively. If p_t is a vector of I(1) variables, the left hand side and first k-1 terms on the right hand side of equation (8) are stationary or I(0), and the first term on the right is a linear combination of first order variables which, given the assumption of the error term, must also be I(0) i.e. $\Pi P_{t-1} \sim I(0)$.

As such Shahidur (2002), noted that the hypothesis of co integration is formulated as a reduced rank of β written as $H(r) : \Pi = \alpha\beta$, where r is the rank of Π that determines how many linear combinations of P_t are stationary, α and β are n x r matrices of full rank. If $r = n$, then the variables are stationary in levels but if $r = 0$, then no linear combination of p_t is stationary.

There are three methods for testing reduced rank of β as reported by Shahidur (2002) and Clemens Lutz *et. al.*, (2006) which are maximum eigenvalue test, expressed as λ_{max} test, and trace test. The null hypothesis is that there is r co integrating vectors and is represented as:

$$(9) \qquad H_0 : \lambda_i = 0 \qquad i = r + 1.....,n$$

Where λ_i is a measure of the strength of correlation between the co integrating relationships for $j = 1.....r$. The maximal-eigenvalue ($\lambda_{\max}$) statistics tests the null hypothesis that there is at most r co integrating vectors against the alternative of r+1 and is given by:

$$(10) \quad \lambda_{\max} = -T \log(1 - \tilde{\lambda}_{r+1}) \qquad r = 0, 1, 2, ..., n-1$$

Where T is the sample size, and $(1 - \tilde{\lambda}_{r+1})$ is the max-eigenvalue estimate.

The trace statistic tests the null hypothesis of r co integrating vectors against a general alternative hypothesis of more than r co integrating vectors and is computed as:

$$(11) \quad \lambda_{trace} = -T \sum_{i=r+1}^{n} \log(1 - \tilde{\lambda}_i) \quad r = 0, 1, 2, ... , n\text{-}1$$

The first test was the Johansen multiple trace test using Maximum likelihood procedure and maximum Eigenvalue and 95% critical value obtained based on residuals from the co integrating regression equations. This Johansen Multiple trace test was run on the null hypothesis that there was r or fewer co integrating vectors against the alternative hypotheses that there were r+1 and at least r+1 co integrations. This test was applied for the Bivariate and multivariate models. This was followed by searching for co integrating vectors following a method explained by Gonzalez-Rivera *et. al.* (2001) to identify the economic market boundary.

3.3.2   Determination of strength of market integration

3.3.2a. Testing for bivariate co integration

There is a tendency for nearby markets to be co integrated. To test for co integration, the two step residual based test developed by Engle and Granger (1987) was used. The first step is the OLS regression or co integrating regression of one I(1) price series say y_{it}, on another I(1) price series say x_{jt}, plus a constant as follows;

$$(12) \quad \Delta y_{it} = \alpha_0 - \alpha_1 \left(y_{it-1} - \beta_1 x_{jt-1} \right) + \beta_0 \Delta x_{jt} + \varepsilon_t$$

Where y_{it} is the price in market i at time t, x_{jt} is the price in market j at time t, α_0, α_1, β_0 and β_1 are parameters to be estimated. In equation (12), the current changes in y_{it} are a function of the current changes in x_{jt} and the degree to which the two series are out of their equilibrium in the previous period. β_0 captures the immediate effect of X on Y described as the contemporaneous or short run effect. β_1 is a coefficient that reflects the equilibrium effect of X on Y.

3.3.2b. Testing for Granger causality relationships

In order to seek an insight about how prices are transmitted across markets and the causal relationships among these markets, another test known as Granger causality test was used. This test is an approach for further market integration testing. A time series p_{it} (rice price in market i at time t) is said to granger cause another time series p_{jt} (rice price in market j at time t) if the current and lagged values of p_{it} improve the prediction of p_{jt} (Gujarat, 1995; Todd Keith and Valerien Pede, 2008). The existence of co integration among a set of variables implies Granger causality (Rashid, 2004), which as indicated by Dawson and Dey (2002), follows the Granger causality approach and can be tested within the Johansen's co integration framework by standard Wald tests. Pair wise causal relationships of such series are re-written as follows:

$$(13) \quad \begin{bmatrix} \Delta p_{1t} \\ \Delta p_{2t} \end{bmatrix} = \begin{bmatrix} \alpha_1 \\ \alpha_2 \end{bmatrix} + \sum_{i=1}^{k-1} \begin{bmatrix} \varphi_{i,11} & \varphi_{i,12} \\ \varphi_{i,21} & \varphi_{i,22} \end{bmatrix} \begin{bmatrix} \Delta P_{1t-i} \\ \Delta P_{2t-i} \end{bmatrix} + \begin{bmatrix} \delta_1 \\ \delta_2 \end{bmatrix} \begin{bmatrix} \beta_1 & \beta_2 \end{bmatrix} \begin{bmatrix} P_{1t-k} \\ P_{2t-k} \end{bmatrix} + \begin{bmatrix} \varepsilon_{1t} \\ \varepsilon_{2t} \end{bmatrix}$$

Where subscripts denote markets and all notations remain the same as in equation (8).
Causation can occur in two ways i.e. unidirectional (one way causation) – where shocks in one market affect another market but not the reverse – and bidirectional (two-way causation) where shocks in one individual market are transmitted both ways. The test for causation is a sufficient condition for short run market integration as longer as at least one causal relationship is confirmed (Ian Coxhead, Agnes Rolla and Kwansoo Kim, 2006). Therefore, based on equation (13), four hypotheses of causality were tested after running a vector autoregression for each market pair. The four hypotheses are:

1. Unidirectional causality from p_{it} to p_{jt} if the coefficients β_{it} are statistically different from zero ($\beta_{it} \neq 0$) and the coefficients $\beta_{i,t}$ are not statistically different from zero ($\beta_{i,t} = 0$).

2. Unidirectional causality from p_{jt} to p_{it} if coefficients $\beta_{j,t}$ are not statistically different from zero and the coefficients $\beta_{i,t}$ are statistically different from zero

3. Bilateral causality (both p_{it} and p_{jt} cause each other) if all coefficients α_i, δ_i, $\beta_{i,t}$ and $\beta_{j,t}$ are statistically different from zero.

4. Independent causality (both p_{it} and p_{jt} do not cause each other) if all coefficients α_i, δ_i, $\beta_{j,t}$ and $\beta_{i,t}$ are not statistically different from zero.

3.3.2c Testing for impulse response functions of the market system

From literature, it is well known that short run parameter estimates generated from VAR models (co integration and ECM) are difficult to interpret. Interpretation of dynamic relationship among prices at alternative markets is best pursued through a consideration of impulse response functions (Goodwin and Piggott, 2001). For this reason, most researchers rely on innovation accounting for short run dynamics from multiple time series. As mentioned by Titus O. Awukose (2007), innovation accounting involves investigating how a one time shock is transmitted through the system via the analysis of Forecast Error Variance Decompositions (FEVDs) and Impulse Response Functions (IRFs). Impulse Response Functions are used as a measure of degree of integration and this is because they trace the effect of a shock over time in location *i* on the price in location *j* (Gloria Gonzalez Rivera *et. al*, 2001). Vector Autoregressive (VAR) modeling especially VECM was used to generate impulse response functions to determine whether the degree of responsive of markets to a one time shock in each of the economic market locations. The IRF can be expressed as the difference between two conditional moment profiles (Baum *et. al.*, 1996) as cited by Uchezuba (2005). It is defined in a similar way to standard IRFs, except that one replaces the linear predictor with a conditional expectation (Potter, 1995). It is expressed as:

$$(14) \qquad I_{t+k}\left(v, y_t, y_{t-1}, \ldots\right) = E\left[Y_{t+n} \middle| Y_t = y_t + v, Y_{t-1}, \ldots\right] - E\left[Y_{t+n} \middle| Y_t = y_t, Y_{t-1} = y_{t-1}, \ldots\right]$$

Where I_{t+k} is the impulse response (y_t, y_{t-1}, $---$) are observed data, v is the shock and $E[.]$ is the expectation operator. The impulse is produced by estimating $E[.]$. Based on the relevancy of the IRFs, the method of analyzing co integration among markets ended with market responses due to various shocks or impulses. The results of these tests for analyzing market integration are also tabulated and discussed in next chapter.

CHAPTER FOUR

RESULTS AND DISCUSSIONS

4.1   Status and extent of market integration

Market integration is an important tool for studying the performance of a marketing system. Therefore, to seek a proper understanding of the performance of Uganda's rice markets, Dickey –Fuller tests whose results can explain the status of integration were run on the price series in the markets.   Table 5 displays the results of these tests and compares their output to the critical values at 1%, and 5% levels of significance.

Table 5: ADF test results for unit root on 2000-2006 average weekly prices

Market	Levels			First Difference			Order of Integration	Critical Values	
	Coefficient	t-statistic	No of Lags	Coefficient	t-statistic	No of Lags		1%	5%
Full Sample (2000 - 2006)									
Kampala	-0.016	-0.803	6	-2.258	-9.789	5	I(1)	-3.455	-2.877
Arua	-0.281	-4.911	5				I(0)	-3.455	-2.877
Gulu	-0.033	-2.029	2	-1.272	-14.619	1	I(1)	-3.455	-2.877
Iganga	-0.055	-2.201	3	-1.651	-13.169	2	I(1)	-3.455	-2.877
Jinja	-0.058	-2.219	3	-1.925	-6.212	9	I(1)	-3.455	-2.877
Kabale	-0.051	-1.314	9	-2.704	-10.975	5	I(1)	-3.455	-2.877
Kasese	-0.003	-0.247	3	-1.741	-13.561	2	I(1)	-3.455	-2.877
Lira	-0.075	-2.478	5	-1.724	-10.013	4	I(1)	-3.455	-2.877
Luwero	-0.057	-1.786	6	-2.466	-6.833	5	I(1)	-3.455	-2.877
Masaka	-0.168	-2.894	10				I(0)	-3.455	-2.877
Masindi	-0.205	-3.311	12				I(0)	-3.455	-2.877
Mbale	-0.077	-1.591	8	-3.098	-9.166	7	I(1)	-3.455	-2.877
Mbarara	-0.114	-2.401	8	-2.500	-7.007	10	I(1)	-3.455	-2.877
Rakai	-0.206	-4.081	9				I(0)	-3.455	-2.877
Soroti	-0.096	-2.140	5	-2.956	-5.213	11	I(1)	-3.455	-2.877
Tororo	-0.427	-6.645	3				I(0)	-3.455	-2.877

Note: The 1% and 5% critical values for levels and first differences are -3.455 and -2.877 respectively.

Table 5 shows results of the Augmented Dickey-Fuller test unit roots for each price series with indicating the order of integration for each test variable. The hypothesis tests are based upon the comparison of calculated statistics with the critical MacKinnon (1991) statistics.

When the full sample prices at level were tested using the ADF test, all markets were non stationary except Arua, Tororo, and Rakai which were exceptionally stationary at level at 1% critical level and Masaka, Masindi too at 5% significance level. This means that these five markets did not share the same trend with the rest of the markets for this traded commodity and information. For Arua, it is probably because it is a regional market that experiences reasonable customer demand from South Sudan and part of North-East Democratic Republic of Congo. Shahidur (2002) while investigating the maize market integration observed this similar dissimilar trend of Arua market, and he suggested that this could have been due to insecurity problems that used to be experienced there in addition to being a border town engaged in trade with neighboring countries. A similar effect of cross border trade could be a reason for Rakai market to fail to share the same trend with the other rice markets besides having joined NAADS program after two years since its inception. The failure of Tororo, Masaka and Masindi markets to share the same trend could be probably due varying degrees of rice production, self sufficiency and hence marketing levels. For instance Tororo district is one of the major rice producing areas presenting a relatively more sufficient market that acts as perhaps a price setter making it have an independent trend while Masaka and Masindi are not self sufficient. Based on the results obtained from this test at level, the same test was conducted for the market prices at first difference and the results show that unit roots are rejected for all prices at the 5% significance level confirming that these markets were integrated at zero order. Overall, it is summarized that among the sixteen markets, the majority price series (69%) are integrated of order one or I (1) at p≤5.

It is important to note that market integration is influenced by many factors. For this reason, two of the major factors that originated from the government policy framework i.e. Plan for Modernization of Agriculture which is the National Agricultural Advisory Services, NAADS program that is implemented country wide. This program and the telecommunication and roads infrastructure development factors were considered very influential in enhancing market integration. However, the two programs began in late 1990's with the effect of privatization of the telecommunication industry taking route by licensing of Celtel – Uganda in 1996 and Mobile Telecommunication Network (MTN) Uganda Ltd in 1998 (UCC, 2000). While these were being commissioned, liberalization of agricultural markets was also done (Kosteki, 1997). These were later followed by enactment of the NAADS Act in 2001 and immediate implementation of the program. As seen from the results of DF tests, the rice markets are integrated under the period of consideration (2000- 2006).

Against this background, this study sought to establish the effect of the two said programs in the early 2000's (2000 to 2003) and their effect in the mid 2000's i.e. 2004 to 2006 on the rice market integration in the country. The data was therefore divided into two samples (2000 to 2003) and 2004 to 2006 as per how the programs were expanding. ADF tests were conducted on each data set and the table that follows depicts the results of the test on integration of markets in the pre NAADS-expansion phase (2000 to 2003).

Table 6: Dickey-Fuller test for unit root by ADF on 2000 -2003 weekly rice prices

Market	Pre-Expansion Period (2000 - 2003)								
	Prices at level			Prices at first difference			Order of integration	Critical values	
	Coefficient	t-statistic	No. of lags	Coefficient	t-statistic	No. of lags		1%	5%
Kampala	-0.123	-1.989	6	-2.475	-7.312	5	I(1)	-3.484	-2.885
Arua	-0.383	-4.490	6				I(0)	-3.484	-2.885
Gulu	-0.105	-2.561	2	-1.479	-12.175	1	I(1)	-3.484	-2.885
Iganga	-0.242	-3.915	1				I(0)	-3.484	-2.885
Jinja	-0.212	-3.541	2				I(0)	-3.484	-2.885
Kabale	-0.349	-3.308	5				I(0)	-3.484	-2.885
Kasese	-0.190	-2.995	3				I(0)	-3.484	-2.885
Lira	-0.247	-3.931	3				I(0)	-3.484	-2.885
Luwero	-0.227	-3.092	4				I(0)	-3.484	-2.885
Masaka	-0.236	-3.219	3				I(0)	-3.484	-2.885
Masindi	-0.215	-3.035	3				I(0)	-3.484	-2.885
Mbale	-0.109	-2.602	2	-1.358	-11.076	1	I(1)	-3.484	-2.885
Mbarara	-0.103	-2.239	2	-1.431	-11.562	1	I(1)	-3.484	-2.885
Rakai	-0.103	-2.102	3	-1.980	-10.515	2	I(1)	-3.484	-2.885
Soroti	-0.168	-2.749	3	-1.826	-10.805	2	I(1)	-3.484	-2.885
Tororo	-0.209	-3.273	3				I(0)	-3.484	-2.885

Note: The 1% and 5% critical values for prices at level and first differences are -3.484 and -2.885 respectively.

From the results (Table 6) few markets; Kampala, Gulu, Mbale, Mbarara, Soroti and Rakai were integrated of order one at level. The rest of the markets were integrated in their first difference i.e. to the zero order. It is probable that since during this period, the NAADS program had not expanded to cover most districts the marketers in these major regional markets were using the improved telecommunication infrastructure to share traded commodity and information on rice product at a higher level than for the rest of the markets. In addition, Gulu, Mbale and Soroti are major rice production centres that could have served as supply centres for Rakai and Mbarara with Kampala as pivotal market for the transactions.

For the remaining markets that were not integrated, this could be explained as having less of the traded commodity even communication had improved as revealed by level of telephony growth during this period. Further, research on rice in Uganda indicates that the Eastern and Northern regions are the major rice producing regions (Odogola, 2006) and central and western regions as less rice producing regions at least before the introduction of Nerica rice in 2004 (Kijima, 2008).   Masaka, Masindi and Kasese markets are located in these low rice production zones that probably made them to be most likely deficit markets. Since these three districts joined NAADS program after 2003 (Appendix I), perhaps one can say that the promotion of rice (especially upland rice) in these districts led to relatively lesser production levels that could have led them to differ from a common trend with other markets.

Goletti *et. al.*, (1995) indicated that there are many reasons for no integration of a market or markets in relation to other markets. Some of the reasons include high transport costs to the dominant markets that can be due to poor roads or high road distances of travel by trucks. Probably due this critical distance, the integration was only evident among major regional trading towns and Kampala dominant market where large scale traders with necessary capital to meet the high transport costs are premised. The failed integration of Arua from the rest of the markets is linked to its geographical location and nature of prevailing business with Southern Sudan. However, as Shahidur (2004) studied the maize market in Uganda, he found Arua price series not sharing a common trend with other markets; he attributed political instability that has disturbed northern Uganda for long as another major factor, which under this period still holds as far as this market is concerned.

Table 7: ADF test results for unit roots on weekly prices 2004 -2006

| Post-Expansion Period (2004 - 2006) | | | | | | | | |
| Market | Levels | | | First Difference | | | Order of Integration | Critical Values | |
	Coefficient	t-statistic	No of Lags	Coefficient	t-statistic	No of Lags		1%	5%
Kampala	-0.055	-1.515	3	-1.515	-8.247	2	I(1)	-3.498	-2.888
Arua	-0.301	-3.412	2				I(0)	-3.500	-2.888
Gulu	-0.106	-2.756	1	-1.105	-8.694	1	I(1)	-3.500	-2.888
Iganga	-0.040	-1.545	1	-1.119	-8.802	1	I(1)	-3.500	-2.888
Jinja	-0.063	-1.608	10	-1.325	-3.151	9	I(1)	-3.500	-2.888
Kabale	-0.105	-1.999	4	-1.738	-8.391	3	I(1)	-3.500	-2.888
Kasese	-0.017	-0.932	3	-1.162	-6.812	2	I(1)	-3.500	-2.888
Lira	-0.046	-0.869	12	-3.018	-5.269	11	I(1)	-3.500	-2.888
Luwero	-0.354	-2.256	4	-1.592	-5.955	3	I(1)	-3.500	-2.888
Masaka	-0.114	-1.216	10	-2.217	-8.585	2	I(1)	-3.500	-2.888
Masindi	-0.153	-2.044	3	-2.005	-8.676	2	I(1)	-3.500	-2.888
Mbale	-0.080	-1.022	8	-3.297	-5.978	7	I(1)	-3.500	-2.888
Mbarara	-0.122	-1.621	8	-2.451	-6.205	7	I(1)	-3.500	-2.888
Rakai	-0.132	-1.652	9	-2.709	-5.896	8	I(1)	-3.500	-2.888
Soroti	-0.033	-0.444	9	-2.933	-7.750	4	I(1)	-3.500	-2.888
Tororo	-0.463	-4.484	3				I(0)	-3.500	-2.888

Note: The 1% and 5% critical values for levels at 4 lags are -3.730 and -2.992 respectively. The 1% and 5% critical values for first differences are -3.500, -2.888 respectively.

Table 7 shows that all markets in the post expansion phase of NAADS program and other infrastructure improvements are integrated of order one save for Arua and Tororo markets. These results could be attributed to many factors but improvement in the communication infrastructure (Fig. 4-1) can be a reason to justify these results. In the post – expansion phase of the NAADS program, all the district markets under study except Owino-Kampala market had been covered under NAADS program while the level of telephony usage had increased during this period by about two million subscribers (Fig. 4-1). Again, from Table 7 all markets except Arua and Tororo are integrated of order one. This indicates that perhaps the NAADS program through the promotion of rice production and marketing and the growth of telephony industry are among the major factors that have led to the improved spatial integration.

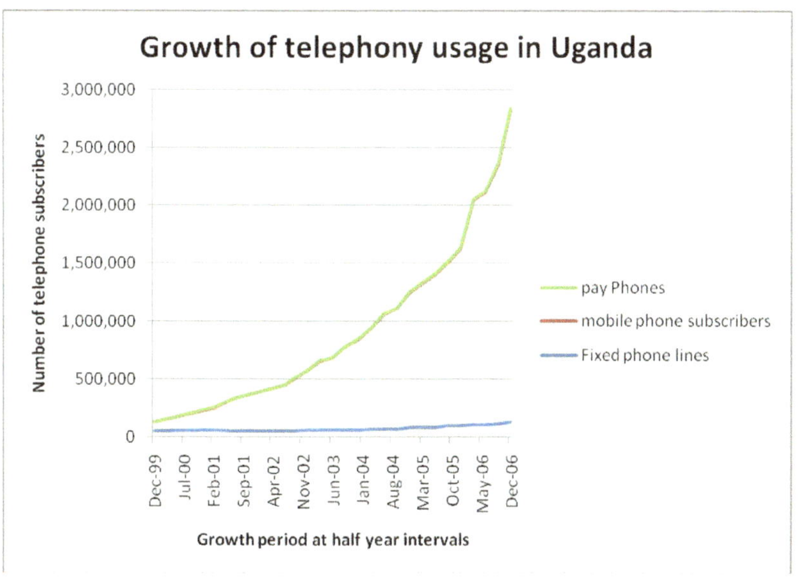

Figure 4 -1: Telephony growth in Uganda over 1999 -2006

Source: © 2005: Ugandaweb

In their study of structural determinants of market integration, Goletti *et. al.*, (1995) pointed out that marketing infrastructure related to transportation and communication is an obvious candidate explanatory variable. They however listed transportation, communication, credit, and storage facilities as what constitutes the marketing infrastructure. Whether the observed integration status is due to credit facilities as reflected by bank density distribution in the study markets is another aspect that will require further research. However, T.S. Hyuha *et. al.*, (2007) cited that Ali and Flinn (1989) and Abdulai and Huffman (2000) reported a negative influence for credit among rice farmers in Pakistan and Northern Ghana, respectively. So, whether this is not the same case with the rice marketing agents is not still a matter of investigation.

4.2 Co integration among price series

After having confirmed that the price series are indeed first difference stationary and integrated, further tests were performed to provide a better understanding of the co movement of rice and rice prices among the markets using bivariate analysis and Johansen co integration tests. Co integration test is based on the notion that if the log likelihood of unconstrained model that includes the co integrating equations is significantly different from the log likelihood of constrained model that does not include the co integrating equations, we reject the null hypothesis of no co integration.

4.2.1. Testing for co integration by bivariate analysis

In an efficient market system, prices move together, "trade takes place if prices in the importing region equal prices in the exporting region plus the unit transport cost incurred by moving between the two markets" (Ravallion, 1986). This test based on two markets known as bivariate analysis was conducted on the basis that Kampala (Owino) market is central and capital city market that plays a pivotal role in the marketing of the rice in the country. The results from this bivariate analysis are tabulated in appendix IV.

From the analysis, the study overwhelmingly rejects the null hypothesis of zero co integrating vectors for most of the market pairs under analysis. It however failed to reject the alternative hypothesis that there is a single co integrating equation in these series. This was especially true for the full sample of the markets. This implies that these markets are integrated in at least one direction except Owino-Masindi link. Out of the 132 market links considered for the full sample, 64 links are significant. This represents 48.5% level of co integration. This is consistent with the results of the stationarity test. The co integration links are more in Gulu, Lira, Iganga, Jinja, Kabale, Kasese, Luwero and Kampala markets. This is likely to be due to the fact that more production of rice is produced in Gulu, Lira, Luwero and Iganga resulting in surpluses that are traded with possibly deficit markets in western Uganda (Kabale & Kasese) with Kampala being an interlinking market. Mbale and Mbarara are almost segmented with very minimal linkage with the majority of the markets except Masindi. The reason for this is not clear but traditionally Mbarara is in a region where rice is not their major staple food crop instead it is banana (Bagamba, 2007) while Mbale market is only significantly linked with Masindi and Mbarara market yet they are distantly farther apart. RATES (2003) cites Mbale and Masindi among the five major maize growing districts in Uganda. It also mentions networking as the main role of large scale traders who are located in major urban centres. On the basis of this role, this is why Mbarara – Mbale, Soroti-Mbale and Masindi – Mbale as well as these regional markets with Kampala market have high and significant links. It is usual for traders to deal in related products like grains, so there is a probability that maize traders in these predominant maize markets are equally involved in rice grain marketing, hence this integration despite farther distance beyond critical 250km as suggested by Goletti *et. al.* (1995).

In the pre-expansion phase too, the study failed to reject the null hypothesis of no co integration in 22 market links out of the 30 links (appendix II). This means that there is slightly more pair wise integrated links during this phase which is 73% extent of co integration. It should be noted that this very high integration is based on only those markets that were integrated of order one during this period. However, this period had less integrated markets (38%) than the post expansion period of the NAADS program and telecommunication infrastructure development as will be discussed later.

In the post expansion phase (2004-2006), the results show that there is more pair wise degree of integrated links of up to 59% representing 108 out of the 182 market links. This suggests that there was an improvement in the national rice market integration. Eight more markets were brought under integration compared to only six that were integrated before the program expansion. Nonetheless, this signifies high market integration (69%) with less integrated market linkages probably being due to delays in transition of the markets to the new group approach of marketing agricultural outputs as opposed to the previous individualist approach by farmers. It should be noted that the NAADS program promotes group approach of marketing agricultural produce especially of outputs from enterprises promoted under this program. Further, Goletti, Ahmed & Farid (1995) pointed out that longer periods of time series are necessary to more robust results because their results were based on 156 weeks covering three years. Therefore given the fact that this post NAADS expansion covered two years and nine months (2004:01 -2006:09), some unexpected results are acceptable. Although in the post expansion phase of NAADS program 64 districts were covered, compared to 21 districts in the pre-expansion phase (NAADS Sec, 2005), in reality this was not achieved within this very period. Based on this coverage trend, the improvement in co integration is perhaps due to the regional interactions in which 49% of Ugandans share market and other information by word of mouth (UBOS, 2002) a medium could happen with or without the NAADS program. However, the observed integration could be partly due to the over a million increase in telephone subscribers and radio stations in the post expansion phase (Ugandaweb, 2005), which play a big role in information dissemination.

4.2.2 Determination of direction of causality relationship among markets

From the bivariate analysis, very many integrated links were observed but there was need to determine the direction of flow of the traded commodity and information. This is accomplished by performing a Granger causality test by running vector autoregression following equation (12). Like in the ADF tests, the lag lengths is very important in determining the outcome and this was ensured by use of AIC and SIC criteria. This test was based on assessing the significance of β_1 and β_2 as specified in equation (12). The results of this test are presented in Table 8a & 8b covering major urban markets (Kampala, Gulu, Mbale, and Mbarara) and the peripheral markets for the entire dataset 2000- 2006.

Table 8a: Granger causality results for markets in pairs with Kampala and regional markets (2000 – 2006)

Major markets	Causality Test	β	R^2	X^2	$p>\chi^2$	Direction of causality
Kampala	Kampala - Gulu	0.748	0.89	13.525	0.000	Two way causation
	Gulu- Kampala	0.695	0.92	37.904	0.000	
	Kampala - Iganga	0.726	0.80	23.189	0.000	Two way causation
	Iganga - Kampala	0.543	0.74	57.418	0.000	
	Kampala-Jinja	0.821	0.79	8.338	0.004	Two way causation
	Jinja -Kampala	0.673	0.71	30.535	0.000	
	Kampala-Kabale	0.805	0.79	16.460	0.000	Two way causation
	Kabale- Kampala	0.446	0.54	19.287	0.000	
	Kampala-Kasese	0.697	0.80	30.712	0.000	Two way causation
	Kasese-Kampala	0.873	0.90	14.889	0.000	
	Kampala-Lira	0.760	0.80	30.751	0.000	Two way causation
	Lira-Kampala	0.462	0.64	79.202	0.000	
	Kampala-Luwero	0.654	0.81	44.830	0.000	Two way causation
	Luwero-Kampala	0.653	0.66	15.330	0.000	
	Kampala-Mbale	0.881	0.77	1.330	0.249	one way causation
	Mbale-Kampala	0.551	0.32	7.869	0.005	
	Kampala-Mbarara	0.882	0.77	0.795	0.373	One way causation
	Mbarara-Kampala	0.449	0.23	6.933	0.008	
	Kampala-Soroti	0.882	0.89	1.444	0.229	one way causation
	Soroti-Kampala	0.643	0.58	6.510	0.011	
Gulu	Gulu -Iganga	0.750	0.84	32.373	0.000	Two way causation
	Iganga- Gulu	0.672	0.71	20.931	0.000	

	Gulu - Jinja	0.858	0.87	14.840	0.000	Two way causation
	Jinja Gulu	0.772	0.74	10.591	0.001	
	Gulu- Kabale	0.868	0.82	5.501	0.019	Two way causation
	Kabale-Gulu	0.523	0.52	35.820	0.000	
	Gulu-Kasese	0.766	0.79	18.944	0.000	Two way causation
	Kasese-Gulu	0.853	0.89	34.137	0.000	
	Gulu - Lira	0.780	0.83	24.498	0.000	Two way causation
	Lira- Gulu	0.440	0.63	63.883	0.000	
	Gulu-Luwero	0.799	0.87	19.999	0.000	Two way causation
	Luwero-Gulu	0.727	0.69	8.929	0.003	
	Gulu - Mbale	0.935	0.87	1.616	0.204	Independent causation
	Mbale-Gulu	0.625	0.38	1.616	0.203	
	Gulu -Mbarara	0.907	0.82	0.033	0.855	Independent causation
	Mbarara-Gulu	0.584	0.34	1.743	0.187	
	Gulu - Rakai	0.883	0.78	0.248	0.618	Independent causation
	Rakai-Gulu	0.512	0.26	0.599	0.439	
	Gulu-Soroti	0.887	0.78	0.710	0.399	Independent causation
	Soroti- Gulu	0.654	0.41	1.296	0.255	
Mbale	Mbale-Iganga	0.535	0.33	9.649	0.002	Two way causation
	Iganga- Mbale	0.792	0.65	4.857	0.028	
	Mbale- Jinja	0.592	0.39	8.150	0.004	Two way causation
	Jinja-Mbale	0.836	0.74	7.223	0.007	
	Mbale-Kabale	0.605	0.39	5.703	0.017	One way causation
	Kabale-Mbale	0.705	0.52	2.606	0.106	
	Mbale-Kasese	0.558	0.32	6.902	0.009	One way causation
	Kasese-Mbale	0.967	0.88	0.519	0.471	
	Mbale-Lira	-0.355	0.32	7.938	0.005	One way causation
	Lira-Mbale	-0.004	0.49	0.062	0.803	
	Mbale -Luwero	-0.120	0.31	2.660	0.103	Independent causation
	Luwero-Mbale	0.035	0.64	2.191	0.139	
	Mbale-Mbarara	0.239	0.32	5.449	0.020	One way causation
	Mbarara-Mbale	0.105	0.22	1.073	0.300	
	Mbale-Soroti	0.537	0.35	21.558	0.000	One way causation
	Soroti-Mbale	-0.158	0.41	2.755	0.097	
Mbarara	Mbarara- Iganga	0.563	0.35	6.408	0.011	Two way causation

Iganga-Mbarara	-0.203	0.70	7.552	0.006	
Mbarara- Jinja	0.554	0.35	7.251	0.007	Two way causation
Jinja-Mbarara	-0.186	0.69	14.109	0.000	
Mbarara- Kabale	0.582	0.34	1.141	0.285	Independent causation
Kabale-Mbarara	-0.135	0.47	2.527	0.112	
Mbarara-Kasese	0.582	0.34	3.616	0.057	Independent causation
Kasese-Mbarara	-0.176	0.90	0.941	0.332	
Mbarara- Lira	0.565	0.32	7.938	0.005	One way causation
Lira-Mbarara	0.708	0.49	0.062	0.803	
Mbarara- Luwero	0.469	0.22	1.972	0.160	Independent causation
Luwero-Mbarara	-0.105	0.64	0.778	0.378	
Mbarara-Rakai	0.375	0.32	5.449	0.020	One way causation
Rakai-Mbarara	0.380	0.22	1.073	0.300	
Mbarara-Soroti	0.127	0.35	21.558	0.000	One way causation
Soroti-Mbarara	0.811	0.41	2.755	0.097	

Causality is significance at p-value of 0.05.

As tabulated above, in the entire study period (2000-06), it was revealed that out the 36 market links with the major regional and capital markets, seventeen were two way with seven being with Kampala market, followed by Gulu market (six) while Mbale and Mbarara each had two two-way causations. The summary of these causalities Table 8b shows that Kampala is the dominant market as expected followed by Mbale, Gulu and Mbarara regional markets respectively.

Table 8b: Summary of Granger causalities in full sample (2000 -2006)

Causality type	Kampala	Gulu	Mbale	Mbarara
Two-way	7	6	2	2
One-way	3	-	5	3
Independent	-	4	1	3

During the pre-expansion phase of the program, the results indicate that out the fourteen causality tests, the spatial integration in the majority (eight) markets, the causalities are unidirectional with only one link being bidirectional. Besides the segmented markets that were stationary at level (table 6) and independent as per causality tests, the tests revealed existence of 64% co integration among the $I(1)$ markets in this period (Table 8c).

Table 8c: Causality results for markets in pairs with Kampala and regional markets

(2000 to 2003)

Major markets	Causality test	β	R^2	X^2	$p>\chi^2$	*Direction of causality*
Kampala	Kampala-Gulu	0.420	0.55	7.655	0.006	One way causation
	Gulu-Kampala	0.151	0.30	1.031	0.310	
	Kampala-Mbale	0.365	0.51	20.458	0.000	One way causation
	Mbale-Kampala	0.346	0.34	1.286	0.257	
	Kampala-Mbarara	0.510	0.27	0.646	0.421	Independent causation
	Mbarara-Kampala	0.109	0.48	3.508	0.061	
	Kampala-Rakai	0.431	0.31	10.870	0.001	One way causation
	Rakai-Kampala	0.208	0.53	1.412	0.235	
	Kampala-Soroti	0.419	0.30	7.226	0.007	Two way causation
	Soroti-Kampala	0.221	0.35	10.387	0.001	
Gulu	Gulu-Mbale	0.776	0.63	2.389	0.122	Independent causation
	Mbale-Gulu	0.103	0.63	0.090	0.765	
	Gulu-Mbarara	0.792	0.63	0.155	0.694	Independent causation
	Mbarara-Gulu	-0.048	0.58	0.180	0.672	
	Gulu -Rakai	0.715	0.55	1.773	0.183	Independent causation
	Rakai-Gulu	0.086	0.54	3.692	0.055	
	Gulu-Soroti	0.686	0.56	5.580	0.018	One way causation
	Soroti-Gulu	0.189	0.32	2.426	0.119	
Mbale	Mbale- Mbarara	0.807	0.63	0.181	0.671	One way causation
	Mbarara-Mbale	-0.038	0.60	9.620	0.002	
	Mbale-Rakai	0.681	0.51	1.378	0.240	One way causation
	Rakai-Mbale	0.264	0.56	14.556	0.000	
	Mbale-Soroti	0.755	0.63	1.304	0.253	One way causation
	Soroti-Mbale	0.402	0.46	31.281	0.000	
Mbarara	Mbarara-Rakai	0.700	0.47	0.493	0.483	Independent causation
	Rakai-Mbarara	-0.049	0.53	0.180	0.672	
	Mbarara-Soroti	0.748	0.59	4.062	0.044	One way causation
	Soroti-Mbarara	0.135	0.37	1.382	0.240	

Causality is significance at p-value of ≤ 0.05.

One way or unidirectional causality implies that only one of the two coefficients is significant i.e. either β_1 or $\beta_2 \neq 0$. For instance from the results (for 2000 – 2003) indicate the Kampala rice market granger causes Gulu prices but not the other way round. This result is interpreted to mean that while Kampala is a capital market, it is only Soroti, which is self sufficient in that at one moment within this period; the rice commodity could be traded in both directions. On the other hand, the remaining markets remained deficit markets which perhaps only relied on Kampala for rice supply while some markets remained indifferent e.g. Kampala - Mbarara in terms of communication.

In the post expansion period, there were 18 bidirectional market links out of the 44 constructed causality tests (Table 8d) and this is about 41%. This is also a positive evidence for improved market performance, in terms of bilateral spatial integration within the market pairs if compared with the situation between 2000 and 2003.

Table 8d: Causality results for markets in pairs with Kampala and regional markets (2004 to 2006)

	Causality Test	β	R^2	X^2	$p>\chi^2$	Direction of causality
Kampala	Kampala - Gulu	0.911	0.83	0.037	0.847	One way causation
	Gulu- Kampala	0.139	0.82	6.417	0.011	
	Kampala - Iganga	0.806	0.84	7.287	0.007	Two way causation
	Iganga - Kampala	0.098	0.91	6.755	0.009	
	Kampala-Jinja	0.660	0.73	10.848	0.001	Two way causation
	Jinja -Kampala	0.170	0.79	8.268	0.004	
	Kampala-Kabale	0.761	0.66	3.249	0.071	One way causation
	Kabale- Kampala	0.189	0.41	12.086	0.001	
	Kampala-Kasese	0.749	0.72	6.608	0.010	One way causation
	Kasese-Kampala	0.044	0.91	3.048	0.081	
	Kampala-Lira	0.680	0.68	9.107	0.003	Two way causation
	Lira-Kampala	0.165	0.51	29.696	0.000	
	Kampala-Luwero	0.816	0.66	0.167	0.683	Independent causation
	Luwero-Kampala	-0.042	0.04	0.102	0.749	
	Kampala-Masaka	0.778	0.72	8.729	0.003	Two way causation
	Masaka-Kampala	-0.465	0.42	10.163	0.001	
	Kampala-Masindi	0.753	0.72	9.702	0.002	Two way causation
	Masindi-Kampala	-0.984	0.47	15.927	0.000	

	Kampala-Mbale	0.757	0.75	11.573	0.001	Two way causation
	Mbale-Kampala	0.416	0.45	21.660	0.000	
	Kampala-Mbarara	0.772	0.72	5.967	0.015	Two way causation
	Mbarara-Kampala	0.448	0.42	17.062	0.000	
	Kampala-Rakai	0.737	0.67	5.614	0.018	Two way causation
	Rakai-Kampala	0.303	0.34	20.450	0.000	
	Kampala-Soroti	0.736	0.73	10.437	0.001	Two way causation
	Soroti-Kampala	0.397	0.47	26.542	0.000	
Gulu	Gulu -Iganga	0.805	0.82	7.917	0.005	One way causation
	Iganga- Gulu	0.947	0.91	0.006	0.936	
	Gulu - Jinja	0.700	0.68	8.219	0.004	One way causation
	Jinja Gulu	0.833	0.74	0.425	0.514	
	Gulu- Kabale	0.604	0.45	2.833	0.092	One way causation
	Kabale-Gulu	0.539	0.38	6.057	0.014	
	Gulu -Kasese	0.892	0.81	0.207	0.649	Independent causation
	Kasese-Gulu	0.006	0.96	0.289	0.591	
	Gulu - Lira	0.724	0.68	7.101	0.008	One way causation
	Lira- Gulu	0.024	0.68	1.802	0.180	
	Gulu-Luwero	0.628	0.44	0.719	0.396	Independent causation
	Luwero-Gulu	-0.025	0.04	0.059	0.809	
	Gulu - Mbale	0.891	0.81	0.104	0.747	One way causation
	Mbale-Gulu	-0.387	0.47	4.069	0.044	
	Gulu -Mbarara	0.885	0.82	0.570	0.450	Independent causation
	Mbarara-Gulu	-0.231	0.57	2.069	0.150	
	Gulu - Rakai	0.888	0.82	0.320	0.571	Independent causation
	Rakai-Gulu	-0.238	0.52	2.169	0.141	
	Gulu-Soroti	0.899	0.81	0.024	0.876	One way causation
	Soroti- Gulu	-0.458	0.53	6.912	0.009	
Mbale	Mbale-Iganga	0.443	0.52	19.210	0.000	Two way causation
	Iganga- Mbale	-0.028	0.92	8.415	0.004	
	Mbale- Jinja	0.484	0.40	9.269	0.002	Two way causation
	Jinja-Mbale	-0.089	0.77	17.611	0.000	
	Mbale-Kabale	0.422	0.29	5.641	0.018	Two way causation
	Kabale-Mbale	-0.049	0.39	8.420	0.004	
	Mbale-Kasese	0.452	0.38	13.542	0.000	One way causation

	Kasese-Mbale	-0.007	0.91	0.964	0.326	
	Mbale-Lira	0.593	0.80	15.717	0.003	Two way causation
	Lira-Mbale	0.415	0.60	32.198	0.000	
	Mbale -Luwero	0.548	0.47	18.068	0.000	Two way causation
	Luwero-Mbale	0.303	0.48	54.767	0.000	
	Mbale-Masaka	0.595	0.47	15.467	0.000	Two way causation
	Masaka-Mbale	0.364	0.34	28.003	0.000	
	Mbale-Masindi	0.440	0.26	0.398	0.528	Independent causation
	Masindi-Mbale	0.440	0.24	0.325	0.569	
	Mbale-Mbarara	0.221	0.28	4.148	0.042	One way causation
	Mbarara-Mbale	0.397	0.20	0.149	0.699	
	Mbale-Rakai	0.255	0.36	9.607	0.002	One way causation
	Rakai-Mbale	0.460	0.22	0.019	0.891	
	Mbale-Soroti	-0.126	0.33	14.235	0.000	Two way causation
	Soroti-Mbale	1.006	0.41	4.904	0.027	
Mbarara	Mbarara- Iganga	0.644	0.59	8.305	0.004	Two way causation
	Iganga-Mbarara	0.891	0.92	11.095	0.001	
	Mbarara- Jinja	0.518	0.41	8.112	0.004	Two way causation
	Jinja-Mbarara	0.758	0.76	16.625	0.000	
	Mbarara- Kabale	0.412	0.21	1.966	0.161	One way causation
	Kabale-Mbarara	0.528	0.38	6.312	0.012	
	Mbarara-Kasese	0.386	0.25	8.262	0.004	One way causation
	Kasese-Mbarara	0.926	0.88	3.145	0.076	
	Mbarara- Lira	0.653	0.63	21.231	0.000	One way causation
	Lira-Mbarara	0.834	0.73	3.362	0.067	
	Mbarara- Luwero	0.450	0.21	1.931	0.165	Independent causation
	Luwero-Mbarara	0.275	0.05	0.730	0.393	
	Mbarara-Rakai	0.329	0.20	0.443	0.506	Independent causation
	Rakai-Mbarara	0.378	0.24	0.405	0.524	
	Mbarara-Soroti	-0.012	0.27	13.025	0.000	One way causation
	Soroti-Mbarara	0.752	0.39	0.805	0.370	

The Granger causations are significant at p = 0.05 critical value.

During this phase, with the exception of segmented links, there are 34 out of the 44 (77.3%) market causality test pairs with either bidirectional or unidirectional causality. This implies that there was a remarkable improvement in the rice market performance. In 2005, David Uchezuba got similar granger causality results while studying South African fresh apple produce markets.

4.3   Results and discussion of multivariate analysis

A bivariate analysis of integration is conducted to identify existence of equilibrium relationships between market pairs over both the medium and long term (K. Weber and D. Lee, 2006). The weakness with this method is that besides its limitation to market pair integration, it does enable one to identify the boundaries of the economic market. Therefore, to provide a further understanding into the level of co integration, another method of testing for co integration known as the Johansen multivariate co integration was employed to the data set. This method enables the researcher to find out the strength of integration among regional markets in the country. This is achieved because unlike the bivariate analysis which excludes or has misspecification problem of the relevant error correction terms and with less explanatory power (R^2) (Gloria Gonzalez- Rivera & Stephan Helfand, 2001), Johansen multivariate analysis enables one to include all the error correction terms and get a better explanation of the extent of market interdependence. The results of this analysis are presented hereafter, in table 9.

Trace Test Ho rank=r	Full Sample (2000 - 2006)		Pre-Expansion Period (2000 - 2003)		Post Expansion Period (2004 - 2006)	
	Trace Statistic	Critical Values (5%)	Trace Statistic	Critical Values (5%)	Trace Statistic	Critical Values (5%)
$r = 0$	527.90	277.71	148.58	94.15	615.89	277.71
$r \leq 1$	407.52	233.13	94.14	68.52	464.60	233.13
$r \leq 2$	323.74	192.89	48.26	47.21	341.18	192.89
$r \leq 3$	251.43	156.00	26.07*	29.68	224.24	156.00
$r \leq 4$	184.43	124.24	10.72	15.41	147.06	124.24
$r \leq 5$	134.87	94.15	4.25	3.76	110.45	94.15
$r \leq 6$	89.55	68.52			79.23	68.52
$r \leq 7$	49.51	47.21			54.46	47.21
$r \leq 8$	25.23*	29.68			31.18	29.68
$r \leq 9$	10.18	15.41			14.68*	15.41
$r \leq 10$	2.30	3.76			0.79	3.76

*Fails to reject a null hypothesis of at least $r = n$ co integrating vectors, at 5% significance level

As tabulated above, when the multivariate analysis was performed, the test rejected the null hypothesis of zero co integrating relation in all the samples and indicating seven co integrating relations in the full set. In comparison however, there were two relations in the pre-expansion phase against eight co integrating relations in the post expansion phase (Table 9). These results are somewhat consistent with earlier findings using bivariate and granger causality analysis in which improved degrees of integration were obtained in terms of integrated markets in the post expansion phase. This is also consistent with the research *apriori* expectation. By this approach, similar results were obtained by K. Weber and D. Lee (2006) and Titus O. Awokuse (2007) on U.S. and China's rice markets in markets were integrated with continuity.

4.3.1   Determination of co integrating market sets

A further test to get the picture of the extent of rice market integration was done through a sequential search for $n-1$ co integrating vectors following the method of Gonzalez-Rivera *et. al.* (2001) and Shahidur (2004). The Johansen's λ_{trace} test results for this sequential search are presented in Table 10. For each set of markets, the null hypothesis of $r = n - 1$ was tested against the alternative that $r < n - 2$, where r is the co integrating vector and n is the number of markets.

Table 10: Results of Johansen LR test for number of co integrating vectors

Markets	r	Trace Statistics	Critical Values
FULL SAMPLE (2000 - 2006)			
Kampala + Gulu	0	40.81	15.41
	1	3.26	3.76
Kampala + Gulu + Iganga	1	35.11	15.41
	2	2.70	3.76
Kampala + Gulu + Iganga + Jinja	2	27.09	15.41
	3	2.33	3.76
Kampala + Gulu + Iganga + Jinja + Kabale	3	26.82	15.41
	4	2.09	3.76
Kampala + Gulu + Iganga + Jinja + Kabale + Lira	4	26.34	15.41
	5	1.38	3.76
Kampala + Gulu + Iganga + Jinja + Kabale + Lira + Mbale	5	17.45	15.41
	6	1.20	3.76
Kampala + Gulu + Iganga + Jinja + Kabale + Lira +Mbale +Mbarara	6	18.12	15.41
	7	1.25	3.76
Kampala + Gulu + Iganga + Jinja + Kabale + Lira +Mbale +Mbarara +Soroti	7	15.91	15.41
	8	1.19	3.76
Kampala + Gulu + Iganga + Jinja + Kabale + Lira +Mbale +Mbarara +Soroti +**Luwero**	8	10.10	15.41
	9	2.99	3.76
Kampala + Gulu + Iganga + Jinja + Kabale + Lira +Mbale +Mbarara +Soroti +**Luwero** +**Kasese**	9	10.18	15.41
	10	2.30	3.76
Pre-expansion period (2000 - 2003)			
Kampala + Gulu	0	37.95	15.41
	1	7.68	3.76
Kampala + Gulu +Mbale	1	16.40	15.41
	2	5.15	3.76
Kampala + Gulu +Mbale +Soroti	2	16.46	15.41
	3	4.91	3.76
Kampala + Gulu +Mbale +Soroti +**Mbarara**	2	11.96	15.41
	3	5.22	3.76
Kampala + Gulu +Mbale +Soroti +**Rakai**	2	15.23	15.41
	3	3.83	3.76
Post Expansion Period (2004 - 2006)			
Kampala + Gulu	0	20.78	15.41

	1	5.06	3.76
Kampala + Gulu + Iganga	1	19.46	15.41
	2	2.78	3.76
Kampala + Gulu + Iganga + Jinja	2	18.95	15.41
	3	2.69	3.76
Kampala + Gulu + Iganga + Jinja +Lira	3	18.27	15.41
	4	1.99	3.76
Kampala + Gulu + Iganga + Jinja +Lira +Rakai	4	17.68	15.41
	5	1.56	3.76
Kampala + Gulu + Iganga + Jinja +Lira +Rakai+ Soroti	5	16.69	15.41
	6	1.13	3.76
Kampala + Gulu + Iganga + Jinja +Lira +Rakai+ Soroti +Mbale	6	16.99	15.41
	7	1.17	3.76
Kampala + Gulu + Iganga + Jinja +Lira +Rakai + Soroti +Mbale +Kabale	7	15.29	15.41
	8	1.13	3.76
Kampala + Gulu + Iganga + Jinja +Lira +Rakai + Soroti +Mbale +Kabale +Masaka	8	14.94	15.41
	9	0.86	3.76
Kampala + Gulu + Iganga + Jinja +Lira +Rakai + Soroti +Mbale +Kabale +Masaka +Masindi	9	14.77	15.41
	10	0.79	3.76

Note: The critical values for rejection of a null hypothesis of $n-2$ are 20.04, 6.65 and 15.41, 3.76 for $P \leq 0.01$ and $P \leq 0.05$ respectively.

As tabulated above, for the full sample the results illustrate that the hypothesis of $n-2$ (i.e., $r = 0$) co integrating vectors is clearly rejected for the first set of markets. However, when Luwero, Masindi, Mbarara, Mbale and Soroti are added, the study fails to reject the null hypothesis of $n-2$ co integrating vectors at 5% level of significance for full sample price series. The same holds when Masaka and Kasese price series are added. It is noted that even if there were sixteen markets, this method revealed that Kampala, Gulu, Iganga, Jinja, Kabale, Lira, Mbale, Mbarara and Soroti markets were co integrated as showed by the search for co integrating vectors in the above table. These nine locations represent the economic market for Uganda's rice industrial market, which is implied by the results of multivariate analysis in the full sample. These results lend support to those obtained in the Granger causality test in Table 8. The rest of the markets were not co integrated.

The integration of Kabale market with the northern and eastern markets is acceptable because Goletti *et. al.*, (1995) explained the role of production levels of the commodity under consideration in influencing the market integration in that it divides the markets into surplus and deficit regions. From available research most of Uganda's rice is produced from eastern and northern Uganda. This result may be taken to mean that Kabale is deficit market, which is not self-sufficient in this product while Gulu, Lira, and Iganga are surplus markets onto which this western market relies for domestic supplies through Kampala, and Jinja markets hence the integration.

In the pre-NAADS expansion phase, there are two significant co integrating relations observed among Kampala, Gulu, and Mbale and Soroti markets. This signifies the less integration among the markets during this period in which sharing the traded commodity and information was restricted to only major urban markets. During this period many markets were not integrated except for these four mentioned including Rakai and Mbarara (Table 10). The remaining ten markets were not included in the co integrating vector search because they were $I(0)$ at level. Among the six integrated markets in this period until Mbarara series was added to the four markets in the co integrating relation, the n-1 co integrating vector hypothesis could not be rejected. When Mbarara was dropped and Rakai added, this did not alter the results. The failure of Mbarara and Rakai to be co integrated with these four markets is perhaps because of other factors that influence market integration. The dissimilarity in the rice production is one of the major factors that could be considered to contribute to varying co integration degrees. For instance Mbarara is among the four major banana growing districts (Bushenyi, Masaka, Mbarara and Mpigi) (Gold *et. al.*, 2002) hence rice is not among its major marketable agricultural outputs.

When analyzed from 2004 to 2006, the price series showed eight significant co integrating relations which were obtained among eleven markets excluding Kasese, Luwero and Mbarara. This is in comparison to the two co integrating relations in the pre–expansion phase. This shows that there was significant improvement in the performance of rice markets in terms of spatial integration supporting the picture that had been painted by the ADF, bivariate and multivariate analyses. During this phase, it is observed that more markets are trading the product and trade information as an economic market composed of Gulu, Jinja, Lira, Rakai, Soroti, Iganga and Kampala as the pivotal market followed by Mbale as revealed by the Granger causality tests (Table 8). Again during the search for the continuous market during this period, the hypothesis of n-1 co integrating vector could not be rejected until Kasese, Luwero and Mbarara markets were added. This means to some extent that the western Uganda markets are not major players in the rice market even as integration improves.

4.3.2 Determination of dynamic adjustments by Impulse Response Functions

It does not suffice to study and identify markets that have long run relationships i.e. with co-movement of prices of the market products. This is because the marketing function is supposed to perform various functions to satisfy the market participants (society). Michael Haines (1998) reports that ignorance of one category of marketers in one area due to perhaps lack of timely information about profits being made by marketing agencies of a given product in another region may result in monopoly over the sell of that product in the concerned market. Therefore, knowledge of how fast certain markets adjust to long run equilibria in the short run is as well required for stakeholders to play their roles in the marketing channels. High market adjustment speed is better for consumers because it deters intermediaries from exploiting them through high pricing.

Although we may take high adjustment rates to be an indicator of the efficiency of the marketing system, Goletti, Ahmed and Farid (1995) however dispute this reasoning and postulate that it is not always true that a high speed of adjustment is an indicator of efficiency, for it could be an indicator of the flexibility mechanism in the system. In Uganda rice production areas are mostly known to be the Northern, eastern parts of the country (Mukiibi, 2001; Imanywoha, 2001) while the western, and central regions of the country are known for other staples other than rice. Therefore, this study sought to analyze the behavior of various   markets in terms response to price signals that emanate from urban markets with reference to Kampala as a dominant market.

Through use of Impulse Response Function analysis the data was subjected Vector autoregression followed by IRFs. Impulse response analysis is employed to investigate the mechanism of shocks i.e. the concept gives information on the impact of price shocks and the way in which shocks are transmitted among market prices. Thus, impulse response functions simulate the effect of a shock in one time-series on itself and another time-series in a system over time. This implies that impulse response functions are examined to determine how quickly prices at one location adapted to shocks in prices at another location prior to and following the identified structural changes for rice markets in Uganda.

It is vital in the impulse response analysis to note that the nonstationarity of price data as well as the error correction properties have a major influence to price shocks. These shocks may elicit responses that are temporary, such that there is a return to the initial time path of the variables, or permanent, causing a persistent shift in the time path (Uchezuba, 2005). Thus, it is necessary to get a better understanding of the dynamic price interrelationships through estimates of an impulse response function. As such, to take care of this phenomenon this study adjusted the last observation in the time series for the Kampala market by one-half standard deviation to represent positive and negative shocks. Responses of these markets in terms of weeks were studied and table 11 summarizes the various levels of variation of responses of the various rice markets to shocks from Kampala market.

Table 11: Results of Kampala impulse response functions on other markets

Market Links	Weeks taken for complete response to occur	
	Pre-Expansion	Post - Expansion
Kampala – Gulu	17	16
Kampala – Mbale	16	15
Kampala – Mbarara	17	16
Kampala – Rakai	14	14
Kampala – Soroti	15	15
Average	15.8	15.2

Note that other markets which were I(0) in either Pre or Post Expansion Period were not included in this analysis to provide a better comparison of the responses.

The impulse response functions for rice markets between Kampala and Gulu as reported in Table 11 and depicted in Figure 4-2 show that 17 weeks were required for Gulu to respond completely to price shocks in Kampala market during the pre period while it took Soroti 15 weeks during post period to respond completely to shocks in Kampala. For Rakai and Soroti market responses (which joined NAADS with Mbarara in similar period) to shocks in Kampala markets (Figure 4-3) do not show any improvement in reduction adjustment period to equilibrium in either phase. To the contrary, Mbarara, which used to take seventeen weeks to completely return to equilibrium, now, requires sixteen weeks in the post expansion period (Table 11 & appendix II). This implies that other factors are involved in influencing the behavior of markets in space and time.

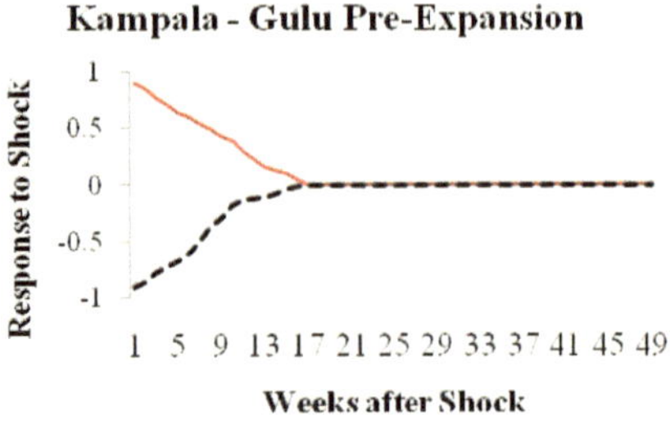

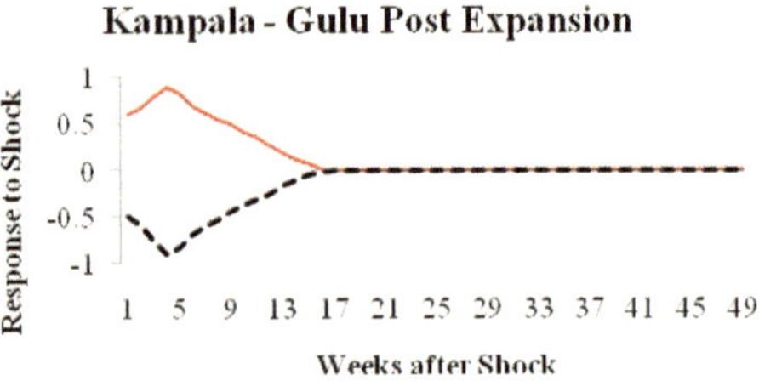

Figure 4-2: Impulse Response Functions of Gulu against Kampala rice market shocks

These results show a relative decline in the period of adjustments during the post period. In general the responses of other auxiliary markets under study are elucidated in Appendix 1I. The results from the IRFs indicate that on average it took 15.6 weeks for auxiliary markets to respond to positive and negative price shocks in the Kampala rice markets. In the post expansion period, a slight improvement was observed whereby auxiliary markets took 15.2 weeks following positive or negative shocks in Kampala after which, the markets restore themselves to equilibria with varying degrees as reflected in the impulse-response graphs.

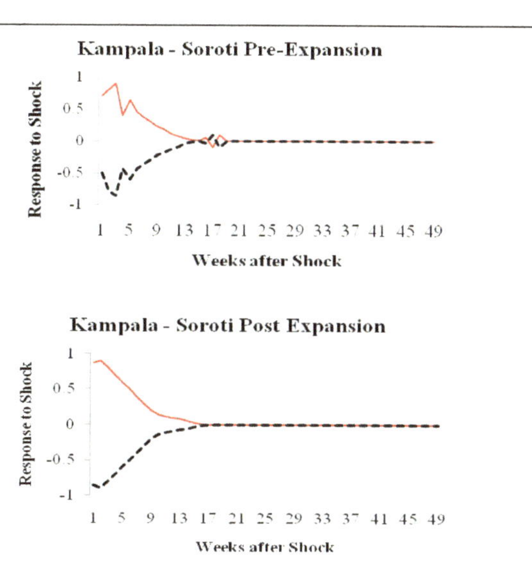

Figure 4-3: Impulse Response Functions for Kampala - Soroti rice markets

The positive and negative shocks converge to equilibrium and do not show any tendency to deviate from equilibrium in the long run. These responses are consistent with long-run market integration. It should, however, be noted that these responses are conditional on the size of the shock due to asymmetric price relationships among the markets.

CHAPTER FIVE

5.0   SUMMARY, CONCLUSIONS AND RECOMMENDATIONS

5.1   Summary

Formulation of market – enhancing policies requires better understanding if how the market of a given commodity functions. Aggregate market performance is better understood by studying the levels of market integration which exist that later is affected by transaction costs in the value chain. In this context, this study was done to assess state of the rice market integration in Uganda following expansion of NAADS program and telecommunication infrastructure since early 2000. It was arranged and conducted to determine whether the rice markets were nationally or regionally integrated by application of various tests of integration and co integration. Below is the tabular summary of results of the various tests for the pre and post expansion periods indicating the percentage changes realized due to the two programs that were considered.

Table 12: Summary results of various tests on integration status in the two program Phases

Comparative test results summary between pre and post expansion NAADS/telecommunication infrastructure programs					
Indicative test	Full sample	pre-program expanded phase	Post- program expanded phase	% age change	P value
STATUS OF INTEGRATION					
1. ADF tests					
I(1) markets observed	12	6.0	14	57.1%	0.02018**
I(0) markets identified	4	10.0	2	80.0%	0.00114**
Total number of markets	16	16	16		
2. Bivariate analysis					
Total market links	132	30.0	182	83.5%	0.00114**
Significant market links	64	22.0	108	79.6%	0.00114**
3. Granger causations to Kampala					
Two way causations	17	1.0	18	94.4%	0.00114**
One way causations	11	8.0	16	50.0%	0.02018**
Independent causations	8	5	8	37.5%	0.14686*

STRENGTH OF INTEGRATION					
4. Multivariate co integrating relations identified	7	2.0	8	75.0%	0.00114**
5. No. of continuous markets identified	8	4.0	8	50.0%	0.02018**
6. Response period to shocks in weeks		15.5	13	3	0.00003**
7. Mean response periods to Kampala Impulses (IRFs)		15.8	15.2	1	0.02018**
Percentage of tests indicating significant market integration				90.9%	
Percentage of test s indicating non significant market integration				9.1%	

NB: ** means results indicate significant market integration improvement at P=0.05. * means the results indicate non significant improvement in integration at P=0.05. P values were obtained after calculating Z vales using formula in appendix III and a decision taken at P = 0.05 using standard Z distribution table.

5.2   Conclusion

From this research, it can now be concluded as follows; by use of weekly wholesale data between 2000 and 2006 obtained from *FoodNet* project for sixteen markets, the integration and extent of co integration were widely investigated using ADF tests. The study was expanded to slightly greater details by analyzing the strength the strength of the spatial integration. This was done by application of bivariate analysis. Nonetheless, due to shortfalls of this bivariate method namely misspecification problem of the relevant error correction terms and with less explanatory power, other methods of testing for strength of integration known as Johansen co integration tests were later conducted to establish the integrated markets where trade and information communication occurs among the marketing agents. It was wound up by investigation of the long run relationship of the markets through Vector Error Correction Model that was later decomposed into moving average representations of the inter-market price relationships. It was on these relationships that impulse response functions were created to evaluate the impact of shocks to prices in one dominant market that could occur within any of the locations (markets) in the economic market and the likely time it would take for shocks to be transmitted and completely eliminated from alternative markets to restore equilibrium. The study examined the behavior of these markets in terms of their price series to determine the impact of structural changes of Uganda's economic policies with reference to the NAADS program and telecommunications infrastructure development between the above periods.

Various co integration tests were conducted and implications from the results drawn in terms of market performance. The co integration results have shown that compared to when the NAADS program and telecommunication infrastructure had expanded by over 50% in terms of national coverage, there was a remarkable improvement in terms of the rice spatial market integration with more markets being economically integrated (table 10 & 12). However, while some markets like Arua and Tororo did not share a common trend with all markets in the entire study period, many others (Iganga, Jinja, Kabale, Kasese, Lira, Luwero, Masaka and Masindi) did not share a common trend with rest of the markets in pre-expansion period. However, with continued expansion and implementation of the programs, all markets except Arua and Tororo became integrated with the rest of the markets in post expansion phase.

From the bivariate analysis, it was found that the level of co integration is highest at 77.8% in the full sample while it was moderate in the NAADS program pre-expansion phase. In the post-expansion period the number of significant market links increased by 79% from 22 links before expansion to 108 links between 2004 and 2006. This improved co integration is supported by results from a multivariate analysis that showed that there was more rice market co integration across the country in the post expansion resulting in up to eight co integrating relations compared to two in the pre-expansion period.

Further analysis of the strength of co integration by sequential search for co integrating vectors was consistent with these results with its final output significantly identifying eight markets; Kampala, Gulu, Iganga, Jinja, Lira, Rakai, Soroti and Mbale in one continuous market in post expanded phase against only Kampala, Gulu, Mbale and Soroti in the 2000 to 2003 period. This means that there was a higher co movement of the traded commodity and trade information on this commodity as was further showed by a relatively high mean adjustment of markets in the short run to the persistent market equilibrium.

The study ended with performing impulse response analyses with Kampala as the source of impulses and results from the IRFs analysis manifested that on average it took 14 to 17 weeks for auxiliary markets to return to equilibrium after any positive and or negative price shocks in the Kampala rice market. In the post expansion period, a further improvement was observed whereby auxiliary markets took 13 to 16 weeks following positive or negative shocks in Kampala after which, the markets restore themselves to equilibria with varying degrees.

Different measures of spatial market integration respond differently to the same structural factors (Goletti et. al. 1995) on which integration and co integration depend but all the various measures employed herein have indicated significant and positive responses. They have shown that Uganda's rice markets are highly integrated and more interdependence of markets in the expanded communication infrastructure than before they had expanded. There is also a growing rice economic market within the country which requires strengthening by supportive factors to marketing. It is surmised that based on these findings, the NAADS program together with the telecommunications growth in the country other factors notwithstanding has really contributed to improved performance particularly the integration of the rice markets in Uganda.

5.3   Recommendations

The Plan Modernization of Agriculture framework was put in place to achieve many objectives including improving efficiency of marketing agricultural commodities. Since the results from this study have revealed that a reasonable improvement in spatial market integration has occurred for the rice market over the study period, there is critical need for operationalization of all other pillars of the PMA especially agro-processing and marketing and rural finance systems to realize significant spatial market integration. The NAADS program should incorporate the provision timely agricultural and marketing information to enhance the current state of integration.

The study has revealed that some markets are by geographical location and environmental endowment, market leaders in this rice markets and constitute the rice economic market. This has made them dominant markets cases in point are Gulu, Iganga, Jinja and Mbale (besides Kampala) markets whose relevance has been paramount in all the phases. Based on this, it is envisaged that southwestern Uganda provides a reliable domestic consumer market for the rice produced by the rest of the country. Government should therefore support optimum rice production in the north and eastern parts of Uganda that are able to sufficiently supply not only for western Uganda but also for the East African region. Government should consider giving more support to such markets to not only provide enough outputs to the rest of the country but also the East African regional market.

Findings from this study have revealed remarkable improvement in spatial rice market integration, but since NAADS program has been promoting production of other crop enterprises, similar studies should be conducted to compare findings to these results. This will help to find out whether such products share the same upcoming economic market.

Leguminous crop enterprise products will be good for consideration for purposes of ensuring recommended crop rotation and soil conservation practices.

Findings from this study have revealed remarkable improvement in spatial rice market integration; similar studies should be conducted for other cereal crop grains to compare findings to these results. This will bring out a broader picture of the performance the cereal markets in Uganda. This will serve as a basis for improving the efficiency of these food crop output markets.

The study was conducted for a period of six years and nine months; this provides a slightly detailed picture of the behavior of this market in terms of integration. However, more robust results are obtained from data for longer periods as was observed earlier by Goletti *et. al.*, (1995) and in addition many other transport and marketing infrastructure are responsible for market integration. Further research based on a time range and consideration of factors necessary for market integration to provide more information on the contribution of the NAADS program and other factors in this aspect of market performance is recommended.

REFERENCES

Abdulai, A. and Huffman, W. 2000. Structural adjustment and economic efficiency of Rice Farmers in Northern Ghana. *Economic Development and Cultural Change* 504-519.

A. R. Ochollah, M.W. Ogenga-Latigo and E.N.B. Nsubuga, 1997. Impact of upland rice cultivation on Crop Choice and income of farmers in Gulu and Bundibugyo districts. **In:** *African Crop Science Conference Proceedings* Volume 3 Part 3 of 3, pp 1407 – 1411.

Agricultural Policy Committee, APC, 1993. Economics of Crops, Bank of Uganda records. *American Journal of Agricultural Economics* 71 (3): 661-669.

Ardeni, P.G. 1989. Does the law of one price really hold for commodity prices? **In**

Arize A., 1995. "The effect of exchange rate volatility on US exports. An empirical investigation," *Southern Economic Journal* 62: 34 – 43.

Asche F., H. Bremnes and Cathy Wessells, 1999. Product aggregation, market integration and relationship between prices: An application to world Salmon market. *American Journal of Agricultural Economics.* Vol.81 568 – 581.

Balke N.S. and Fomby T.B. 1997. "Threshold co integration" International Economic Review No. 78 pp 625-45.

Barrett Christopher B., 1996. "Market analysis methods, Are our enriched Toolkits well suited to Enlivened markets?" *American Journal of Agricultural Economics.* Vol.78: 825-829.

Baulch B. 1997. Transfer costs, spatial arbitrage and testing for food market integration. *American Journal of Agricultural Economics.* Vol.79 (20):477-487.

Baum C.F., Barkoulas J.T. and Caglayan M., 2001. Non linear adjustment of purchasing power parity in the post –Breton era. Journal of International Money and Finance, Vol. 20: 379 -399.

Bresnahan T.F., 1989. Empirical studies of industries with market power', in Schmalense, R. and Willig Robert (Eds): Handbook of industrial Organization. Vol. II Ch. 17. North Holland, Amsterdam.

Clemens Lutz, Cornelis Praagman and Luu Thanh Duc Hai, 2006. Rice market integration in the Mekong River Delta: The transition to market rules in the domestic food market in Vietnam. *The European Bank for Reconstruction and development, Blackwell Publishing Ltd Oxford, UK. Economics of Transition 16 (3) 517-546.*

Cooper, Donald R., and Emory C. William. 1995. *Business Research Methods*. Irwin, 5th ed., pp. 681. [HD30.4 Coo HMLMS, ISBN 0-2561-3777-3]

Dawson P.J. and P.K. Dey, 2002. Testing for the Law of One Price: Rice market integration in Bangladesh. Journal of International Development 14: 473 – 484.

Enders W. and C.W.J Granger, 1998. Unit root test and asymmetric adjustment with an example using the term structure interest rates. *Journal of Business and Economic statistics,* Vol. 16 (3) 304 -311.

Enders Walter, 1995. Applied econometric Time Series, USA. John Wiley & Sons Inc.

Engel R.F.C. and C.W.J Granger, 1987. Co integration and error correlation,

FAO/GIEWS, 2006. Review of food and crops in Uganda. Special Report.

Gardener L. Bruce, 1975. The farm-Retail Spread in competitive Industry. In *American Journal of Agriculture Economics* Vol. 57 No.3 New York.

Gittinger J, P., Leslie J. & Hoistings C. 1987 Food Policy: "Integrating Supply Distribution" Ed series in Economic Development. World Bank John Hopkins University press Baltimore London.

Gold C. S., A Kiggundu, A. M. K. Abera and D Karamura, 2002. Diversity, distribution and farmer preference of *Musa* cultivars in Uganda. IITA, Eastern and Southern Regional Centre, Kampala, Uganda.

Goletti Francesco and Eleni Christina-Tsigas, 1995. Analyzing market integration. Prices, products and people, Analyzing Agricultural markets in Developing countries, International Potato Centre, Lima, Peru.

Goletti Francesco, Rausuddin Ahamed and Nasser Farid, 1995: Structural Determinants of Market integration. A case of rice markets in Bangladesh. *The Developing Economies,* XXXIII – 2.

Gonzalez Gloria Rivera and S.M. Helfand, 2001. The extent, pattern and degree of market integration: A multivariate approach for the Brazilian rice market. *American Journal of Agricultural Economics* 83 (3) August 2001, pp. 576- 592.

Goodwin J. Barry and Schroeder C. Ted (1990), Co integration tests and spatial price linkages in regional cattle markets. **In** America *Journal of Agricultural Economics,* Vol. 73 (2) 1991 pp 452 – 462.

Goodwin K. Barry and Nicholas E. Piggott, (2001). Spatial market integration in the presence of threshold effects. American Journal of Agricultural Economics Vol. 83(2): 302 -317.

Granger Clive W. J, 2004. Time series analysis, Co integration and Application. *The American Economic Review* Volume 94 No. 3 pp 421 –425.

Gujarat, N. Damodar, 2003. Basic Econometrics 4th edition McGraw Hill United States Military Academy, West Point.

Ian Coxhead, Agnes Rola and Kwansoo Kim, _______. Chapter 3. Philippine Development Strategies, Price policies and National markets; Where is the linkages in Lantapan?

Imanywoha J.B., 2001, Rice. **In** *Agriculture in Uganda,* crops Volume 2 pg 70. Fountain publishers/CTA/NARO.

Istoqomah, Manfred Zeller and Stehan von Cramon-Taubadel, 2005. Volatility and integration of rice markets in Java, Indonesia: A comparative Analysis before and after Trade Liberalization. *Conference on International Agricultural Research for Development,* Stuttgart – Hohenheim, October 11- 13, 2005.

Jane Ininda, 2005. Charting the future of rice in sub-Saharan Africa, Alliance of Green Revolution in Africa. http//www.WARDA.org/newsletter/no8/htm.

Jason R.V. Franken and Joel L. Parcell, 2003. "Market integration: Case studies of change." Proceedings of NCR-134 conference on applied commodity price analysis, Forecasting and market risk management. St. Louis MO http//www. farmdoc.uiuc.edu/nccc134.

Jesus Gonzalo and Jean–Yves Pitarakis, 2000. Lag length estimation in large Dimensional systems. Universidad Carlos III de Madrid and University of Reading.

Johansen S. 1988. "Statistical analysis of co integration vectors" *Journal of Economic Dynamic Control.* 12:231 -54.

K. Weber and D. Lee, 2006. Spatial Price Integration in U.S. and Mexican Rice Markets. *A paper prepared for presentation at the international Association for agricultural Economists Conference*, Gold Coast, Australia, August 12-18, 2006.

Kohls & Uhl, 1990. "Marketing of Agricultural Products." Prentice Hall, New Jersey.

Kosteki M. Michael and Bernard M. Hoeckman, 1997. The Political Economy of World Trading System: From GATT to W.T.O. Oxford University press.

Kyomuhendo Jacqueline, 2003. Returns to Resources in Rice production among small scale farmers in Pallisa District, Uganda. An MABM thesis, Makerere University Kampala.

Le Dang Trung, Tran Ngo Minh Tam, Bob Baulch and Henrik Hansen, 2007. Testing for Food market integration: A study of the Vietnamese paddy market. Depocen Working Paper Series No. 2007/11. http://www.depocenwp.org

Luke Keele & Suzanna De Boef, 2004. Not just for co integration: Error Correction Models with stationary data. Department of political science and international relations. Nuffield College and Oxford University, UK

MAAIF/MFPED, 2000. Ministry of Agriculture Animal Industry and Fisheries/ministry of Finance Planning and Economic Development. Plan for Modernization of Agriculture. Government Strategy for Eradicating Poverty, Kampala Uganda.

MAAIF/UBOS, 2002. Ministry of Agriculture, Animal Industry & Fisheries/Uganda Bureau of Statistics. Population Census preliminary report.

Mackinnon J. G., 1991. Critical values for co integration tests' in Engle R. F. and Granger C. W. J. eds. Long run Economic Relationships: readings in Co integration, Oxford University Press, pp 267-276.

MFPED/APC, 1997. Operationalization of the medium Term Plan for Modernization of Agriculture, 1997/98 to 2001/2002.

MFPED/Uganda Bureau of Statistics, 2005. The 2002 Uganda Population Housing Census, main Report. The Republic of Uganda.

Michael Haines, 1998. Marketing for farm & rural enterprise. Farming Press, UK.

NARS, 2002. Master document of the NARS review task force Draft report, Ministry of Agriculture, Animal Industry and Fisheries, October 2002.

Negassa A, Meyers R and Gabriel Maldhin E., 2003. Analyzing the grain market efficiency in developing countries: Review of existing methods and extending to Parity bound models. Market trade and institutions division. Discussion 63.

Oryokot J. O. E., 2001. Cereals, introduction. **In** *Agriculture in Uganda, Crop.* Vol. 2 Fountain publishers/CTA/NARO.

Potter S.M., 1995. A non linear approach to U.S. GNP. *Journal of Applied Econometrics,* Vol. 16: 109 -125.

Raghaendra Jha, K. V. B. Murthy, Hari K. Nagarajan and Ashok Seth, 1999. Components of the wholesale bid-ask spread and structure of grain markets: the case of rice in India. **In** *Agriculture Economics* 21 (1999) 173 – 189.

Ravallion M. 1986. Testing Market Integration. *American Journal of Agricultural Economics.* Vol. 68: 102 – 109.

Regional Agricultural Trade Intelligence Network, RATIN, 2005. Food Trade Bulletin for East Africa Issue No.28.

Representation, estimation and testing. *Econometrica* 55(2): 251-76.

Schmalense, R. and Willig Robert 1989: Handbook of industrial Organization. North Holland.

Shahidur Rashid, (2002). Liberalization, access to information and spatial integration of Ugandan maize markets: A dynamic multivariate analysis. Annual workshop on policies for improved Land Management in Uganda Project, *International Food Policy research Institute, UK.*

Stifel D. Laurence, 1975. Imperfect competition in a vertical market network. The case of rubber in Thailand. **In** *American Journal of Agricultural Economics.* Vol. 57 number 4.

Stock, Jame H and Watson, Mark W, 1987. Interpreting Evidence on Money-Income causality. NBER working paper series, Vol. 2228.

T.S. Hyuha, B. Bashaasha, E. Nkonya and D. Kraybill, 2007. Analysis of Profit Inefficiency in Rice Production in Eastern and Northern Uganda

T.S. Hyuha, E. N. Sabitti and E. Hisal, 2005. Impact of rice production on food security and women in Uganda. **In** *African Crop Science Proceedings* Vol. 5. pp 632 -637.

Theingi Myint and Siegfried Bauer, 2005. Rice Market integration in Myanmar. *Conference on International Agricultural Research for Development,* Stuttgart – Hohenheim, October 11- 13, 2005.

Thomas L. Cox and Jean-Paul Chavas, 1998. An interregional analysis of price discrimination and domestic policy reform in the U.S. dairy sectors. **In** *American Journal of Agricultural Economics* 83 (1) February 2001.

Titus O. Awokuse, 2007. Market reforms, spatial price dynamics and China's rice market integration: A causal analysis with directed acyclic graphs. *Journal of Agricultural and Resource Economics* 32(1) 58-76.

Todd H. Kwethe and Valerien Pede, 2008. Regional Housing Price Cycles; a Spatio-temporal analysis using U. S. State level data. Working Paper # 08 - 14. Department of Agricultural Economics, Purdue University,

Tomek, William G. and Kenneth L. Robinson, **1990**. "Introduction." Chapter 1 in Agricultural. Product Prices. Third Edition, Ithaca: Cornell University Press, "Co integration: Implications for the Market Efficiencies.

Uchezuba Ifeanyu David, 2005. Measuring market integration for apples on the South African Fresh produce market: A threshold Error Correction Model. An M sc. Thesis, University of Free Town, Bloemfontein, South Africa.

WARDA, 2005. Charting the future of rice in sub-Saharan Africa, Africa Rice Centre.

WARDA, 2007. 2007 Africa Rice trends, Africa rice trends 5th edition.

Wijaya, H.R., 1985. Women's access to land resources. Some observations from East Javanese rural agriculture. Pages 171 – 185. **In:** *Women in rice farming. Proceedings of a conference on women in rice farming systems.*

Yoko Kijima, 2008. New Technology and Emergence of Markets; evidence from Nerica rice in Uganda. Discussion paper No. 165, Graduate school of International Development, Nagoya University, Japan.

Appendix I: Map of Uganda Showing NAADS Program Expansion from 2001/02 to 2007/08

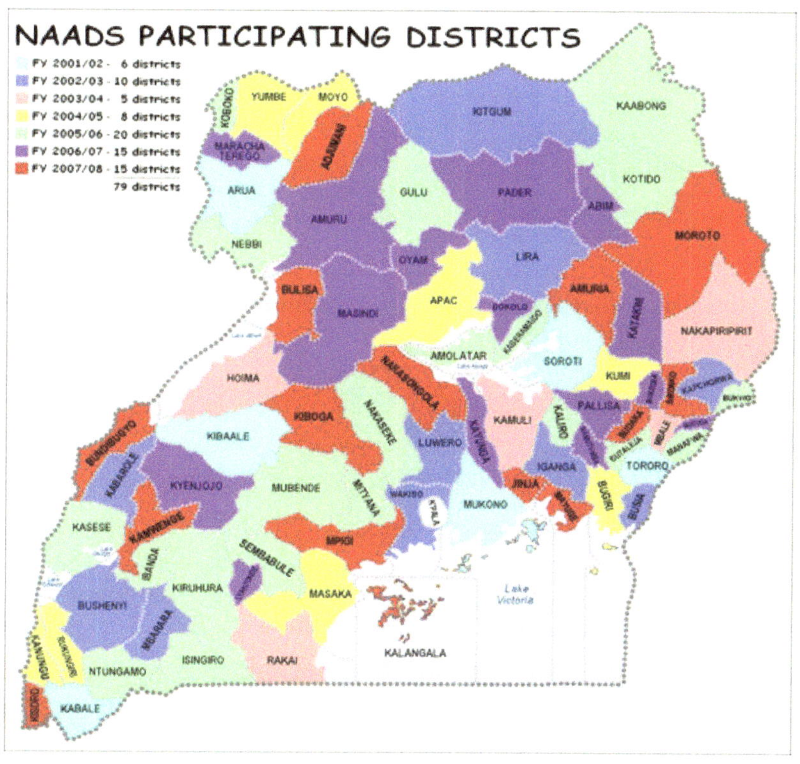

Appendix II: IRF graphs of Kampala market shocks against other markets

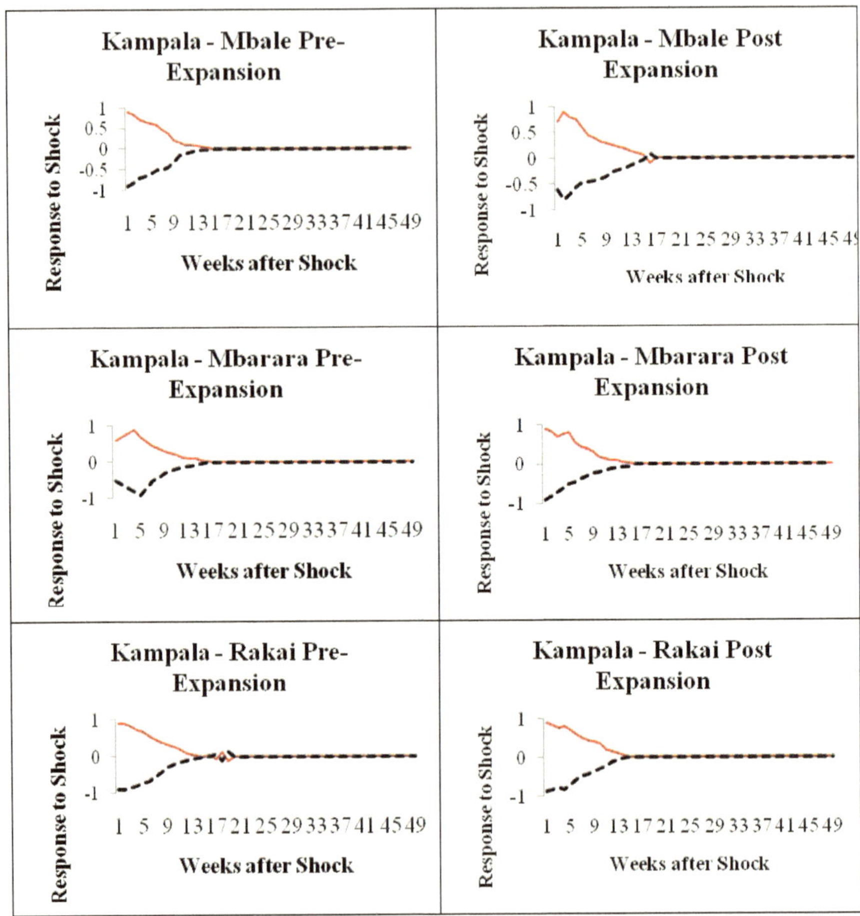

Appendix III: Proportional Difference test for market integration

To test for the significance of market integration, the proportional difference for pre-NAADS program expanded phase and the post expansion period a given co integration parameter estimator, the Z-test was performed as per the following hypothesis and formula:

$$H_0 : P_i = P_j$$

$$H_1 : P_i \neq P_j$$

$$Z = \frac{p_i - p_j}{\sqrt{P_c \dfrac{(1-P_c)}{n_i} + P_c \dfrac{(1-P_c)}{n_j}}}$$

Where $P_c = \left(\dfrac{X_i + X_j}{n_i + n_j} \right)$

And Z = Standard normal cumulative probabilities

     P_i = Proportion of pre- NAADS /telecommunication programs expansion

     P_j = Proportion of post NAADS /telecommunication programs expansion

     P_c = Pooled estimate of P_i and P_j and constant.

     X_i = Number of co integrated markets during the pre-expanded programs Period

     X_j = Number of co integrated markets during the post-expanded program period

     n_i = Number of integrated markets during the pre- program expansion period

     n_j = Number of integrated markets during the post program expanded period.

Appendix IV: Bivariate analysis results for rice market co integration

Market (I,j)	Kampala	Gulu	Iganga	Jinja	Kabale	Kasese	Lira	Luwero	Masaka	Masindi	Mbale	Mbarara	Rakai	Soroti
FULL SAMPLE (2000 - 2006)														
Kampala	0.00	40.81**	36.86**	21.46**	38.29**	19.03*	36.08**	25.16**			5.02	13.70		36.41**
Gulu	40.81**	0.00	38.70**	28.35**	38.28**	15.45*	41.43**	26.74**			7.36	12.33		9.73
Iganga	36.86**	38.70**	0.00	38.61**	39.18**	17.09*	36.05**	23.19**			8.45	13.16		12.01
Jinja	21.46**	28.35**	38.61**	0.00	10.68	12.03	31.87**	21.67**			8.91	12.95		7.54
Kabale	38.29**	38.28**	39.18**	10.68	0.00	41.49**	26.58**	40.82**			7.63	11.51		7.16
Kasese	19.03*	15.45*	17.09*	12.03	41.49**	0.00	22.82**	19.15**			5.55	9.03		6.58
Lira	36.08**	41.43**	36.05**	31.87**	26.58**	22.82**	0.00	36.92**			9.20	12.12		13.93
Luwero	25.16**	26.74**	23.19**	21.67**	40.82**	19.15**	36.92**	0.00			20.74**	25.21**		14.56
Mbale	5.02	7.36	8.45	8.91	7.63	5.55	9.20	20.74**			0.00	27.58**		17.2**
Mbarara	13.70	12.33	13.16	12.95	11.51	9.03	12.12	25.21**			27.58**	0.00		31.19**
Soroti	36.41**	9.73	12.01	7.54	6.58	6.58	13.93	14.56			17.21*	31.19**		0.00
PRE-EXPANSION PERIOD (2000 - 2003)														
Kampala	0.00	37.95**									44.04**	19.34*	21.90**	18.95*
Gulu	37.95**	0.00									17.21*	13.99	13.97	27.80**
Mbale	44.04**	17.21*									0.00	19.77*	14.78	43.91**
Mbarara	19.34*	13.99									19.77*	0.00	10.71	26.58**
Rakai	21.90**	13.97									14.78	10.71	0.00	18.80*
Soroti	33.88**	27.80**									43.91**	26.58**	18.80*	0.00

Market (I,j)	Kampala	Gulu	Iganga	Jinja	Kabale	Kasese	Lira	Luwero	Masaka	Masindi	Mbale	Mbarara	Rakai	Soroti
POST EXPANSION PERIOD (2004 - 2006)														
Kampala	0.00	20.78**	24.23**	24.96**	22.29**	5.85	21.26**	6.59	9.54	27.36**	51.00**	34.59**	37.11**	20.30*
Gulu	20.78**	0.00	19.78*	17.19*	21.28**	9.29	21.56**	11.67	9.76	13.11	21.03**	28.49**	31.81**	19.28*
Iganga	24.23**	19.78*	0.00	16.56*	19.34*	6.56	22.22**	7.42	9.81	30.90**	40.11**	38.84**	41.60**	47.33*
Jinja	24.96**	17.19*	32.46**	0.00	14.21	5.86	32.98**	7.72	10.57	20.85**	30.22**	28.55**	33.58**	34.72**
Kabale	22.29**	21.28**	19.34*	14.21	0.00	23.78**	14.73	10.74	15.36	11.77	6.51	18.96*	37.29**	45.13*
Kasese	5.85	9.29	6.56	5.86	23.78**	0.00	8.98	13.92	5.75	7.16	5.07	21.59**	24.38**	11.3
Lira	21.26**	21.56**	16.70*	32.98**	14.73	8.98	0.00	10.95	17.87*	50.00**	20.73**	18.09*	28.64**	35.28**
Luwero	6.59	11.67	7.42	7.72	10.74	13.92	10.95	0.00	14.81	14.28	11.73	15.65*	11.57	6.9
Masaka	9.54	9.76	9.81	10.57	15.36	5.75	10.95	14.81	0.00	38.20**	20.29**	19.71*	38.77**	19.30
Masindi	27.36**	13.11	18.33*	20.85**	11.77	7.16	50.00**	14.28	7.15	0.00	20.80**	37.85**	13.98	20.85*
Mbale	51.00**	37.18**	40.11**	30.22**	6.51	5.07	20.73**	11.73	20.29**	20.80**	0.00	23.55**	9.49	16.84*
Mbarara	16.71*	28.49**	38.84**	28.55**	18.96*	21.59**	18.09*	15.65*	19.71*	37.85**	23.55**	0.00	11.17	10.7
Rakai	16.42*	31.81**	41.60**	33.58**	37.29**	24.38**	28.64**	11.57	38.77**	8.45	9.49	11.17	0.00	14.4
Soroti	20.30**	36.37**	47.33**	34.72**	45.13**	4.91	35.28**	6.92	19.30*	23.14**	19.53*	10.78	14.45	0.0

The critical values for rejection of a null hypothesis of no co integration are 20.04 and 15.41 for P ≤ 0.01 and P ≤ 0.05 respectively. An integrated link between market I and market j is the one whose Trace statistic is above the critical value. Any value >15.41 but ≤ 20.04 is significant at 5% while any value >20.04 is significant at 1%. ** and *means significant at 1% and 5% significance levels respectively . Arua market is I (0) in the full sample, Iganga, Mbale, Gulu, Lira and Arua in pre-expansion phase and Arua in post expansion phase are I (0).

YOUR KNOWLEDGE HAS VALUE

- We will publish your bachelor's and
  master's thesis, essays and papers

- Your own eBook and book -
  sold worldwide in all relevant shops

- Earn money with each sale

Upload your text at www.GRIN.com
and publish for free